T0312880

Chapman & Hall/CRC FINANCIAL MATHEMATICS SERIES

Model-Free Hedging

A Martingale Optimal Transport Viewpoint

Pierre Henry-Labordère

 CRC Press
Taylor & Francis Group
Boca Raton   London   New York

CRC Press is an imprint of the
Taylor & Francis Group, an **informa** business
A CHAPMAN & HALL BOOK

CHAPMAN & HALL/CRC
Financial Mathematics Series

Aims and scope:
The field of financial mathematics forms an ever-expanding slice of the financial sector. This series aims to capture new developments and summarize what is known over the whole spectrum of this field. It will include a broad range of textbooks, reference works and handbooks that are meant to appeal to both academics and practitioners. The inclusion of numerical code and concrete real-world examples is highly encouraged.

Series Editors

M.A.H. Dempster
Centre for Financial Research
Department of Pure
Mathematics and Statistics
University of Cambridge

Dilip B. Madan
Robert H. Smith School
of Business
University of Maryland

Rama Cont
Department of Mathematics
Imperial College

Published Titles

American-Style Derivatives; Valuation and Computation, *Jerome Detemple*

Analysis, Geometry, and Modeling in Finance: Advanced Methods in Option Pricing, *Pierre Henry-Labordère*

C++ for Financial Mathematics, *John Armstrong*

Commodities, *M. A. H. Dempster and Ke Tang*

Computational Methods in Finance, *Ali Hirsa*

Counterparty Risk and Funding: A Tale of Two Puzzles, *Stéphane Crépey and Tomasz R. Bielecki, With an Introductory Dialogue by Damiano Brigo*

Credit Risk: Models, Derivatives, and Management, *Niklas Wagner*

Engineering BGM, *Alan Brace*

Financial Mathematics: A Comprehensive Treatment, *Giuseppe Campolieti and Roman N. Makarov*

The Financial Mathematics of Market Liquidity: From Optimal Execution to Market Making, *Olivier Guéant*

Financial Modelling with Jump Processes, *Rama Cont and Peter Tankov*

Interest Rate Modeling: Theory and Practice, *Lixin Wu*

Introduction to Credit Risk Modeling, Second Edition, *Christian Bluhm, Ludger Overbeck, and Christoph Wagner*

An Introduction to Exotic Option Pricing, *Peter Buchen*

Introduction to Risk Parity and Budgeting, *Thierry Roncalli*

Introduction to Stochastic Calculus Applied to Finance, Second Edition, *Damien Lamberton and Bernard Lapeyre*

Model-Free Hedging: A Martingale Optimal Transport Viewpoint, *Pierre Henry-Labordère*

Monte Carlo Methods and Models in Finance and Insurance, *Ralf Korn, Elke Korn, and Gerald Kroisandt*

Monte Carlo Simulation with Applications to Finance, *Hui Wang*

Nonlinear Option Pricing, *Julien Guyon and Pierre Henry-Labordère*

Numerical Methods for Finance, *John A. D. Appleby, David C. Edelman, and John J. H. Miller*

Option Valuation: A First Course in Financial Mathematics, *Hugo D. Junghenn*

Portfolio Optimization and Performance Analysis, *Jean-Luc Prigent*

Quantitative Finance: An Object-Oriented Approach in C++, *Erik Schlögl*

Quantitative Fund Management, *M. A. H. Dempster, Georg Pflug, and Gautam Mitra*

Risk Analysis in Finance and Insurance, Second Edition, *Alexander Melnikov*

Robust Libor Modelling and Pricing of Derivative Products, *John Schoenmakers*

Stochastic Finance: An Introduction with Market Examples, *Nicolas Privault*

Stochastic Finance: A Numeraire Approach, *Jan Vecer*

Stochastic Financial Models, *Douglas Kennedy*

Stochastic Processes with Applications to Finance, Second Edition, *Masaaki Kijima*

Stochastic Volatility Modeling, *Lorenzo Bergomi*

Structured Credit Portfolio Analysis, Baskets & CDOs, *Christian Bluhm and Ludger Overbeck*

Understanding Risk: The Theory and Practice of Financial Risk Management, *David Murphy*

Unravelling the Credit Crunch, *David Murphy*

Proposals for the series should be submitted to one of the series editors above or directly to:
CRC Press, Taylor & Francis Group
3 Park Square, Milton Park
Abingdon, Oxfordshire OX14 4RN
UK

CRC Press
Taylor & Francis Group
6000 Broken Sound Parkway NW, Suite 300
Boca Raton, FL 33487-2742

Printed on acid-free paper
Version Date: 20170419

International Standard Book Number-13: 978-1-1380-6223-8 (Hardback)

Visit the Taylor & Francis Web site at
http://www.taylorandfrancis.com

and the CRC Press Web site at
http://www.crcpress.com

Contents

Preface   ix

1 Pricing and hedging without tears   1
  1.1   An insurance viewpoint   1
       1.1.1   Utility preference   2
       1.1.2   Quantile approach   3
  1.2   A trader viewpoint   5
       1.2.1   Super-replication: Linear programming   6
       1.2.2   Arbitrage-free prices and bounds   9
       1.2.3   A worked-out example: The binomial model   11
       1.2.4   Replication paradigm   12
       1.2.5   Geometry of $\mathcal{M}_1$: Extremal points   13
       1.2.6   Mean-variance: Quadratic programming   14
       1.2.7   Utility function: Convex programming   16
       1.2.8   Quantile hedging   16
       1.2.9   Utility indifference price   17
       1.2.10   A worked-out example: The trinomial model   18
  1.3   A cautious trader viewpoint   20

2 Martingale optimal transport   25
  2.1   Optimal transport in a nutshell   25
       2.1.1   Trading T-Vanilla options   25
       2.1.2   Super-replication and Monge–Kantorovich duality . .   26
       2.1.3   Formulation in $\mathbb{R}_+^d$ and multi-dimensional marginals .   31
       2.1.4   Fréchet–Hoeffding solution   31
       2.1.5   Brenier's solution   34
       2.1.6   Axiomatic construction of marginals: Stieltjes moment
             problem   37
       2.1.7   Some symmetries   39
       2.1.8   Robust quantile hedging   41
       2.1.9   Multi-marginals and infinitely-many marginals case . .   42
       2.1.10   Link with Hamilton–Jacobi equation   43
  2.2   Martingale optimal transport   44
       2.2.1   Dual formulation   47
       2.2.2   Link with Hamilton–Jacobi–Bellman equation   51
       2.2.3   A discrete martingale Fréchet–Hoeffding solution . . .   53
       2.2.4   OT versus MOT: A summary   58

	2.2.5	Martingale Brenier's solution	58
	2.2.6	Symmetries in MOT	60
	2.2.7	c-cyclical monotonicity	61
	2.2.8	Martingale McCann's interpolation	61
	2.2.9	Multi-marginals extension	64
	2.2.10	Robust quantile hedging	67
	2.2.11	Model-independent arbitrage	69
	2.2.12	Market frictions	70
2.3		Other optimal solutions	71
2.4		Numerical experiments	76
2.5		Constrained MOT	76
	2.5.1	VIX constraints	78
	2.5.2	Entropy penalty	82
	2.5.3	American options	86

3 Model-independent options — **89**

3.1		Probabilistic setup	89
3.2		Exotic options made of Vanillas: Nice martingales	90
	3.2.1	Variance swaps	90
	3.2.2	Covariance options	92
	3.2.3	Lookback/Barrier options	93
	3.2.4	Options on spot/variance	96
	3.2.5	Options on local time	97
3.3		Timer options	98
	3.3.1	Dirichlet options	98
	3.3.2	Neumann options	100
	3.3.3	Some generalizations	102
	3.3.4	Model-dependence	103
3.4		Ocone's martingales	105
	3.4.1	Lookback/Barrier options	108
	3.4.2	Options on variance	109
	3.4.3	Options on local time	110

4 Continuous-time MOT and Skorokhod embedding — **113**

4.1		Continuous-time MOT and robust hedging	113
	4.1.1	Pathwise integration	113
	4.1.2	Continuous-time MOT	115
4.2		Matching marginals	116
	4.2.1	Bass's construction	117
	4.2.2	Local variance Gamma model	118
	4.2.3	Local volatility model	120
	4.2.4	Local stochastic volatility models and McKean SDEs	121
	4.2.5	Local Lévy's model	122
	4.2.6	Martingale Fréchet–Hoeffding solution	123
4.3		Digression: Matching path-dependent options	128

4.4   Link with Skorokhod embedding problem 129
4.5   A (singular) stochastic control approach 131
4.6   Review of solutions to SEP and its interpretation in mathematical finance   134
    4.6.1   Azéma–Yor solution 134
    4.6.2   Root's solution 146
    4.6.3   Perkins solution   152
    4.6.4   Vallois' solution 156
4.7   Matching marginals through SEP   160
    4.7.1   Through Azéma–Yor 161
    4.7.2   Through Vallois   162
    4.7.3   Optimality in $\mathcal{M}^{c}((\mathbb{P}^{t})_{t \in (0,T]})$ 162
4.8   Martingale inequalities   163
    4.8.1   Doob's inequality revisited   163
    4.8.2   Burkholder–Davis–Gundy inequality 165
    4.8.3   Inequalities on local time   167
4.9   Randomized SEP   168
    4.9.1   Robust pricing with partial information 169
    4.9.2 ρ-mixed SEP 170
    4.9.3   Optimality 172

References   **179**

Index   **189**

Preface

> Ainsi, l'on voit dans les Sciences, tantôt (...) des théories brillantes, mais longtemps inutiles, devenir tout à coup le fondement des applications les plus importantes, et tantôt des applications très simples en apparence, faire naître l'idée de théories abstraites dont on n'avoit pas encore le besoin, diriger vers les théories des travaux des Géomètres, et leur ouvrir une carrière nouvelle.
>
> Nicolas de Condorcet, Rapport sur les déblais et les remblais[1]

This book focuses on the computation of model-independent bounds for exotic options consistent with market prices of liquid instruments such as Vanilla options.

The main problem, when evaluating an exotic option, is how to choose an "appropriate" pricing model, a model being characterized by a martingale measure from the no-arbitrage condition in mathematical finance. "Appropriate" here means that the model allows to capture the main risks of the exotic option under consideration: at-the-money volatility, skew, forward volatility, forward skew,... One may impose that the model is calibrated to a set of (liquid) market instruments or matches some historical levels. Because Vanilla options are liquid, hence the most suitable hedge instruments, the model has to comply with their market prices. In mathematical terms, the marginals of the underlying under the probability measure, for a discrete set of dates, are given. Only a few models such as Dupire's local volatility or (multi-factor) local stochastic volatility models can achieve efficient calibration to Vanilla options.

Here we follow a different route. Instead of postulating a model, we focus on the computation of model-independent bounds consistent with Vanillas, eventually additional instruments such as VIX futures. By duality, we will show that these bounds are attained by some arbitrage-free models. The computation of model-independent bounds for exotic options can then be framed as a constrained optimal transport problem, the so-called martingale optimal transport problem (in short MOT).

In this book, we give an overview of MOT, highlighting the differences between the optimal transport (in short OT) and its martingale counterpart. We explore this topic in the context of mathematical finance. Optimal transport, first introduced by G. Monge in his work "Théorie des déblais et des

---

[1] Histoire de l'Académie royale des sciences avec les mémoires de mathématique et de physique tirés des registres de cette Académie (1781), 34-38.

remblais" (1781), has recently generated wide interest in various mathematical domains as highlighted by the Fields medallist C. Villani (2010). Let us mention the analysis of non-linear (kinetic) partial differential equations arising in statistical physics such as McKean–Vlasov PDE, infinite-dimensional linear programming, mean-field limits, convergence of particle methods and the study of Ricci flows in differential geometry. Despite its numerous ramifications in analysis and probability, OT had not yet attracted the attention of academics/practitioners in financial mathematics. One aspect of this book is to illustrate the use of MOT in mathematical finance.

On a practical side, MOT allows us to better understand payoffs that can be well-hedged by Vanilla options. On a theoretical side, MOT leads to new results in probability and mathematical finance: new solutions to the multi-marginals Skorokhod embedding problem, a new approach for deriving pathwise martingale inequalities, new peacock processes, and robust hedging. The MOT that we introduced in an internal publication at Société Générale in 2009 has attracted much attention: various summer schools and workshops have been devoted to it. MOT is already mature enough to give an overview of this subject, accessible to both practitioners and academics. In this respect, we deliberately leave out technical conditions (mainly in stochastic analysis and convex analysis). Details can be found in (our) research papers. For example, we barely distinguish between local and true martingales. We have deliberately limited the number of references because of the format of the book. We would like to apologize to readers whose papers are not cited.

Additionally, some important aspects of MOT are not covered, such as robust fundamental theorems of asset pricing and c-cyclical monotonicity. However, to inform the readers about these topics, we have included some short sections pointing to some relevant references.

Although our concern is more on applications (in pricing and hedging exotic derivatives) than results in probability, our presentation uses the traditional definition, lemma, proposition, and theorem. In our experience, this allows the reader to better differentiate what is defined and what is proved. In particular, all our theorems are proved (sometimes using simple assumptions or skipping technical details). For people looking for these mathematical details, we have included some references we think can be useful for filling the gaps.

Guide for the reader

In order to guide the reader, here is a general description of the book chapters:

In Chapter 1, we give a quick introduction to mathematical finance for academics not familiar with option pricing. Our approach, mainly restricted to a discrete-time setting, emphasizes the super-replication approach that we will use when introducing MOT. This chapter can be skipped for people already familiar with these notions. We will insist on the model-dependence of our approach.

Chapter 2 focuses on MOT in a discrete-time setting. In order to highlight the main differences with OT, we first present a quick recap of this subject. In particular, we review Monge–Kantorovich duality, its formulation by Hamilton–Jacobi equation, and Fréchet–Hoeffding (Brenier) solution. All these concepts are then extended to the martingale version of OT. Our presentation is illustrated by practical examples in finance.

Chapter 3 deals with model-independent payoffs. They can be perfectly hedged (eventually using static position in Vanillas). Model-independence means that the arbitrage-free price is unique within a large class of models (typically modeled by the set $\mathcal{M}^c$ of continuous martingales). An example of such payoffs is provided by timer options which depend on path-dependent states (invariant by time-change) such as running maximums, realized variances, and local times. These options will pop up naturally in the next chapter. In the last part, we show how to enlarge the class of model-independent payoffs by restricting the class $\mathcal{M}^c$ to the subset of Ocone's martingales.

Chapter 4 considers MOT in the continuous-time case. In finance, this corresponds to a robust superhedging problem when the trader is allowed to dynamically hedge the underlying and to invest in Vanillas. By performing a stochastic time change, this problem will be linked with the so-called Skorokhod embedding problem (in short SEP) which consists in building a stopping time τ such that a (geometric) Brownian motion stopped at that time has the same law as a probability μ: $B_\tau \sim \mu$. In option pricing layman's term, by the Dambis–Dubins–Schwarz time change theorem, it consists of building an arbitrage-free model consistent with T-Vanillas for some maturity T. Then, we show that MOT is an efficient tool for deriving pathwise martingale inequalities. Finally, we will consider a constrained MOT where we impose the value of the spot/volatility correlation - say ρ. This is linked with a randomized version of SEP. This problem interpolates nicely between the replication price within the class of Ocone's martingales (i.e., $\rho = 0$) and our continuous-time MOT (i.e., $\rho = 1$).

I hope that by reading this book you will enjoy my ride in mathematical finance.

Book's audience

This book is mainly designed for quantitative analysts. In particular, no knowledge of OT is needed (this was my case when jumping into the subject). However, for people working in OT, who might want to apply their own research to MOT, Chapter 1 should provide an introduction to option pricing. Our introduction closely follows our discussion on (M)OT in the next chapter and therefore should not be seen as esoteric for someone with only a background in OT. (M)OT is then illustrated by various examples in finance. This should help quantitative analysts (resp. applied mathematicians) to understand (M)OT (resp. option pricing).

Acknowledgments

I would like to thank my collaborators on MOT: M. Beiglböck, J. Claisse, S. De Marco, A. Galichon, G. Guo, J. Obloj, F. Penkner, P. Spoida, X. Tan and N. Touzi. Working together on this subject has been exciting and allows MOT to go far beyond its original purpose. We gratefully acknowledge J. Claisse, H. Guennoun and G. Guo for a careful reading and N. Touzi for his help in writing Section 4.1.1.

An oral presentation of this book was given during the winter school in mathematical finance held at Lunteren (2014). Parts of this book were written when the author was preparing his habilitation thesis (HDR) in applied mathematics [92]. We take the opportunity to again thank our referee committee: B. Bouchard, Y. Brenier, N. El Karoui, P. Friz, D. Talay and N. Touzi.

About the author

Pierre Henry-Labordère works in the Global Markets Quantitative Research team at Société Générale. He holds a Ph.D. in Theoretical Physics from Ecole Normale Supérieure (Paris) and a habilitation thesis in Applied Mathematics from University Paris-Dauphine. More importantly, Pierre has a long-standing experience in tek diving, particularly mixed-gas closed-circuit rebreathers. Pierre is a professor (chargé de cours) at Ecole Polytechnique and a research associate at CMAP (Ecole Polytechnique). He was the recipient of the 2013 "Quant of the Year" award from Risk Magazine and the 2014 Institute Louis Bachelier award for his paper on MOT written in collaboration with M. Beiglböck and F. Penkner from the University of Vienna.

Figure 1: Ready for a ride in mathematical finance?

Chapter 1

Pricing and hedging without tears

Abstract   Let us denote F_T an European payoff depending on the value S_T of an asset at T. The main questions, that we will focus on, are

(A) At which price should we sell this option? Is there a unique price?

(B) After having sold this option, what should we do in order to reduce (i.e., *hedge*) our potential losses?

By working on a discrete-time setting, we will try to answer these two questions without relying on knowledge of stochastic analysis, but by using classical tools in optimization and in probability. In particular, we will formulate the problem of pricing and hedging of derivative products as some linear, quadratic and convex optimization problems. This consists in minimizing an utility function written on the profit and loss wealth value of a delta-hedged portfolio. In particular, we will insist on convex duality from which *risk-neutral* models emerge as *dual* variables. This convex (linear) duality will appear naturally in the next chapter and will be a key tool when we will discuss (M)OT. We will also highlight the model-dependence of our approach.

1.1   An insurance viewpoint

Let us assume that we have sold an option (at $t = 0$) at the price C. An option, with payoff F_T, gives the right to the holder to exercise the payoff at a maturity T. Its gain at T is F_T. For example, $F_T = (S_T - K)^+$ for a call and $F_T = (K - S_T)^+$ for a put where S_T is the asset price at T and K is called the strike. Although in this part we will focus mainly on call options as a simple example, our discussion remains valid for a general (non American) option with payoff $F_T \equiv F(S_T)$. From $t = 0$ up to the maturity T, we do not apply any hedging strategy (see Section 1.2). The cash C is invested (at $t = 0$) into a bank account with a fixed interest rate r. The value of our portfolio at T, composed of only our bank account, is Ce^{rT}. At T, the client exercises his option and our portfolio value becomes finally

$$\pi_T \equiv e^{rT} C - F_T$$

which is equivalent to the discounted portfolio value:

$$e^{-rT}\pi_T = C - e^{-rT}F_T \tag{1.1}$$

How should we fix C?

1.1.1 Utility preference

A first approach is to choose C such that the expectation of π_T vanishes, $\mathbb{E}^{\mathbb{P}^{\text{hist}}}[\pi_T] = 0$, i.e.,

$$C_{\text{ins}} = e^{-rT}\mathbb{E}^{\mathbb{P}^{\text{hist}}}[F_T] \tag{1.2}$$

The expectation is taken under a probability $\mathbb{P}^{\text{hist}}$ that describes the (historical) fluctuations of S_T. More precisely, uncertainty is described by a probability space $(\Omega, \mathcal{F}, \mathbb{P}^{\text{hist}})$ where Ω is the set of possible outcomes, $\mathcal{F}$ is a σ-algebra that describes the set of all possible events (an event is a subset of Ω) and $\mathbb{P}^{\text{hist}}$ is a probability describing the likelihoods of the various events.

Note that S_T is a positive random variable, in principle atomic as S_T take discrete values. Leaving apart the atomic nature of the distribution of S_T, we choose here for the sake of simplicity a log-normal density

$$p(S_T = K) = \frac{1}{\sqrt{2\pi\sigma^2 T}K} \exp -\frac{\left(\ln\frac{K}{S_0} + \frac{1}{2}\sigma^2 T - \mu T\right)^2}{2\sigma^2 T} \tag{1.3}$$

Here S_0 denotes the asset value at $t = 0$ (which is known). This density depends on two parameters: a drift μ and a (log-normal) volatility σ. Note that μT and $\sigma\sqrt{T}$ are dimensionless and satisfies

$$\mathbb{E}^{\mathbb{P}^{\text{hist}}}\left[\frac{S_T}{S_0}\right] = e^{\mu T} \tag{1.4}$$

$$-\frac{2}{T}\mathbb{E}^{\mathbb{P}^{\text{hist}}}\left[\ln\frac{S_T}{S_0}\right] = \sigma^2 - 2\mu \tag{1.5}$$

By direct integration of (1.2) with $F_T = (S_T - K)^+$, we obtain for a call option with strike K and maturity T:

$$C_{\text{ins}}(T, K) = e^{(\mu-r)T}S_0 N(d_1) - Ke^{-rT}N(d_2) \tag{1.6}$$

with

$$d_1 = \frac{1}{\sigma\sqrt{T}}\left(\ln\frac{S_0}{K} + (\mu + \frac{1}{2}\sigma^2)T\right), \quad d_2 = d_1 - \sigma\sqrt{T}$$

$N(x) \equiv \int_{-\infty}^{x} e^{-\frac{y^2}{2}} dy$ is the Gaussian cumulative distribution. Note that this formula depends here on the drift μ (see the scaling $e^{(\mu-r)T}$ and the expression of d_1), a priori *different* from r. It is at least debatable whether the drift is really observable or even statistically measurable. At this point, the price C is strongly model-dependent as it depends on our choice of the density (through the two parameters μ and σ).

If we want to use this formula, we need to estimate statistically the two parameters μ and σ using for example the two relations (1.4, 1.5). Finally, the variance of our portfolio is

$$V \equiv \mathbb{E}^{\mathbb{P}^{\text{hist}}}[\pi_T^2] - \mathbb{E}^{\mathbb{P}^{\text{hist}}}[\pi_T]^2 = \mathbb{E}^{\mathbb{P}^{\text{hist}}}[((S_T - K)^+)^2] - e^{2rT}C_{\text{ins}}(T,K)^2$$

This variance could be large and therefore the seller of the option faces an important risk. As a numerical application, for $r = 5\%$, $\mu = 10\%$, $\sigma = 20\%$ and $S_0 = 100$, the price of a call option with $T = 1$ year and $K/S_0 = 1$ - we say *at-the-money* call option - is from formula (1.6) $C_{\text{ins}} = 13.95$. The standard deviation is $\sqrt{V} = 17.80$, which is quite large (greater than the premium C_{ins}).

REMARK 1.1 Fat tails   In practice, the distribution of log-daily returns for a stock is not well-modeled by a normal probability density, i.e., $\ln \frac{S_{t_{i+1}}}{S_{t_i}} = (\mu - (1/2)\sigma^2)\Delta t + \sigma\sqrt{\Delta t}Z_i$, $(Z_i)_{i=1,\ldots,N} \in N(0,1)$ and independent, $\Delta t = 1$ day, and hence $\ln S_T$ is not well-modeled by a normal density. (Asymmetric) Student's distributions, which have fat tails, are known to fit best the distributions of returns and can therefore be used to price call options from formula (1.2). Note that Student's distributions are not stable by convolutions and therefore one needs to rely on some numerical integration methods.   ▯

This approach can be generalized by introducing an utility function U which is a negative concave function. We set C such that

$$C^* = \operatorname*{argmax}_{C \in \mathbb{R}} \mathbb{E}^{\mathbb{P}^{\text{hist}}}[U(\pi_T)]$$

Taking the derivative with respect to C, we get the first-order condition:

$$\mathbb{E}^{\mathbb{P}^{\text{hist}}}[U'(e^{rT}C^* - F_T)] = 0$$

Our previous formula (1.2) corresponds to take $U(x) = -x^2$.

1.1.2   Quantile approach

An alternative approach is to set C such that our (potential) losses are under controlled: for a fixed probability $p \in (0, 1]$, we choose C such that

$$C_{\text{ins}}(p) = \inf\{C \ : \ \mathbb{P}^{\text{hist}}[\pi_T \geq 0] \geq p\}$$

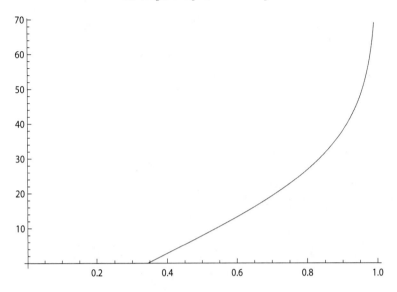

Figure 1.1: $C_{\text{ins}}(p)$ as a function of $p \in (0,1)$ for an at-the-money call option (i.e., $K = 100$). Parameters for the log-normal density: $r = 5\%$, $\mu = 10\%$, $\sigma = 20\%$ and $S_0 = 100$. For reference, the price (1.2) obtained using a log-normal density is 13.95.

Here we control only the likelihoods of losses, not their size. $C = C_{\text{ins}}(p)$ is necessarily nonnegative for nonnegative payoffs and therefore for a call payoff,

$$\mathbb{E}^{\mathbb{P}^{\text{hist}}}[1_{\pi_T \geq 0}] = \mathbb{E}^{\mathbb{P}^{\text{hist}}}[1_{S_T \geq K} 1_{S_T \leq Ce^{rT}+K}] + \mathbb{E}^{\mathbb{P}^{\text{hist}}}[1_{S_T \leq K}]$$
$$= F_{\mathbb{P}^{\text{hist}}}(Ce^{rT}+K)$$

where

$$F_{\mathbb{P}^{\text{hist}}}(x) \equiv \mathbb{E}^{\mathbb{P}^{\text{hist}}}[1_{S_T \leq x}]$$
$$= 1 + e^{rT}\partial_x C_{\text{ins}}(T,x)$$

with $C_{\text{ins}}(T,x)$ defined by (1.6). Finally,

$$C_{\text{ins}}(p) = e^{-rT}(F_{\mathbb{P}^{\text{hist}}}^{-1}(p) - K)^+$$

We have plotted $C_{\text{ins}}(p)$ as a function of p (see Figure 1.1). For $p = 1$, $C_{\text{ins}}(p)$ diverges as expected as $F_T = (S_T - K)^+$ is an unbounded payoff.

In Table 1.1, we summarized our different formulations for the pricing of F_T. Note that the price depends strongly on our choice of $\mathbb{P}^{\text{hist}}$. Statistical studies are therefore needed in order to estimate the right distribution of S_T. We are still far away from the notion of model-independence.

Table 1.1: Summary of our different formulations for the pricing of F_T.

Quadratic utility	$\mathbb{E}^{\mathbb{P}^{\text{hist}}}[\pi_T] = 0$, $C_{\text{ins}} = e^{-rT}\mathbb{E}^{\mathbb{P}^{\text{hist}}}[F_T]$
Utility preference	$C^* \equiv \text{argmax}_C \mathbb{E}^{\mathbb{P}^{\text{hist}}}[U(\pi_T)]$
Quantile	$C_{\text{ins}}(p) \equiv \inf\{C \ : \ \mathbb{P}^{\text{hist}}[\pi_T \geq 0] \geq p\}$

1.2 A trader viewpoint

At this point, the seller of our previous option (with payoff F_T) learns from his management that it can buy and sell the asset on the market. If we denote by H the initial number of assets bought by the trader, the portfolio value at $t = 0$ is

$$\pi_0 = -HS_0 + C$$

HS_0 corresponds to the price of a portfolio consisting of H assets with unit price S_0 at $t = 0$. Note that H could be positive or negative. If $H > 0$, we say that we have a *long position*, this means that the trader has bought H shares. How to achieve $H < 0$ (we say that we have a *short* position) as the trader does not hold the asset at $t = 0$? The trader borrows at $t = 0$ the share from a counterparty and sell it on the market at $t = 0$, generating a profit $-HS_0 > 0$. He gives back the share to the counterparty at the maturity T at the price S_T with a *small* premium, called the *repo*. For the sake of simplicity, we consider here that the repo is null. The portfolio value at T is then

$$\pi_T \equiv (-HS_0 + C)e^{rT} + HS_T - F_T$$

which is equivalent to

$$e^{-rT}\pi_T = H\left(S_T e^{-rT} - S_0\right) + C - e^{-rT}F_T$$

$(-HS_0 + C)e^{rT}$ is the value of our bank account, HS_T the value of H units of our asset (at the price S_T) and F_T the payoff exercised at the maturity T. Note that in comparison with Equation (1.1), we have the additional term $H\left(S_T e^{-rT} - S_0\right)$, generated from our delta-hedging strategy.

In the next section, we focus first on question (A). As in the previous section, the price will be fixed by maximizing an utility function depending of our portfolio value π_T. These different optimizations will be framed as linear, quadratic and convex programming problems. As in the first section, we assume that

Assumption 1 *The random variable S_T is well-modeled by an historical probability $\mathbb{P}^{\text{hist}}$.*

In next chapters, Assumption 1 will be weakened by assuming that $\mathbb{P}^{\mathrm{hist}}$ is uncertain but belongs to a specific class (for example arbitrary distributions with prescribed marginals). For the moment, we don't know how precise should be our knowledge and description of $\mathbb{P}^{\mathrm{hist}}$: should we know its support, its negligible sets, its density, etc? Remember that the density was explicitly required in our insurance's approach - see Formula (1.6) and Table 1.1.

1.2.1 Super-replication: Linear programming

We define *seller's super-replication* price C as

DEFINITION 1.1 Seller's super-replication price

$$C_{\mathrm{sel}} \equiv \inf\{C \ : \ \exists \, H \text{ s.t. } \pi_T \geq 0, \quad \mathbb{P}^{\mathrm{hist}} - \text{a.s.}\} \qquad (1.7)$$

where the abbreviation a.s. means almost surely. C and H are chosen such that the portfolio value at T is nonnegative for all realization of S_T distributed according to the law $\mathbb{P}^{\mathrm{hist}}$. This definition is clearly that of non-adverse risk traders.

Here, note that our definition depends weakly on our modeling assumption only through the negligible sets of $\mathbb{P}^{\mathrm{hist}}$. $\mathbb{P}^{\mathrm{hist}}$ can be replaced by *any* probability $\mathbb{Q}$ equivalent to $\mathbb{P}^{\mathrm{hist}}$ - see Definition 1.2.

For completeness, we recall the definition of equivalent probabilities:

DEFINITION 1.2 Equivalent probabilities $\mathbb{P} \sim \mathbb{Q}$ - *we say $\mathbb{P}$ and $\mathbb{Q}$ are equivalent on a sigma field $\mathcal{F}$ - if $\mathbb{P}$ and $\mathbb{Q}$ have the same negligible sets: $\mathbb{P}(A) = 0$ if and only if $\mathbb{Q}(A) = 0$ for all $A \in \mathcal{F}$.*

For example if $\mathbb{P}^{\mathrm{hist}}$ is an atomic probability supported on the points $(s_i)_{i=1,\ldots,N}$ with probabilities $(p_i \equiv \mathbb{P}^{\mathrm{hist}}(S_T = s_i) \neq 0)_{i=1,\ldots,N}$, $\mathbb{Q} \sim \mathbb{P}^{\mathrm{hist}}$ means that $\mathbb{Q}$ is also an atomic probability supported on the same points $(s_i)_{i=1,\ldots,N}$ with probability $(q_i \neq 0)_{i=1,\ldots,N}$.

C_{sel} defines a so-called *infinite*-dimensional linear programming problem: we need to compute the infimum of a linear cost, i.e., C, with respect to the real variables C and H, subject to an *infinite* number of constraint inequalities, parameterized by S_T. If we assume that $\mathbb{P}^{\mathrm{hist}}$ is an atomic probability supported on N points $(s_i)_{i=1,\ldots,N}$ - this is the case in practice as the unit price is one cent - the super-replication price can be stated as a more conventional finite-dimensional linear programming (written here for $F_T = (S_T - K)^+$)

$$C_{\mathrm{sel}}^N \equiv \inf\{C \ : \ \exists \, H \text{ s.t. } H\left(s_i e^{-rT} - S_0\right) + C - e^{-rT}(s_i - K)^+ \geq 0,$$
$$i = 1,\ldots,N\}$$

C_{sel}^N can then be solved numerically using a simplex algorithm. We will now present a dual formulation (Monge–Kantorovich dual) of C_{sel} that will be fundamental for understanding the notion of arbitrage-free prices and risk-neutral probabilities.

THEOREM 1.1 Seller's super-replication price
Let $F_T \in L^\infty(\mathbb{P}^{\text{hist}})$. Then,

$$C_{\text{sel}} = \sup_{\mathbb{Q} \in \mathcal{M}_1} \mathbb{E}^{\mathbb{Q}}[e^{-rT} F_T] \tag{1.8}$$

with $\mathcal{M}_1 \equiv \{\mathbb{Q} : \mathbb{Q} \sim \mathbb{P}^{\text{hist}}, \quad \mathbb{E}^{\mathbb{Q}}[e^{-rT} S_T] = S_0\}$.

Taking for example for $\mathbb{P}^{\text{hist}}$ a log-normal density with drift (1.3), then a log-normal density with a drift equal to the rate r:

$$\mathbb{Q}(S_T = K) = \frac{1}{\sqrt{2\pi\sigma^2 T} K} \exp - \frac{\left(\ln \frac{K}{S_0} + \frac{1}{2}\sigma^2 T - rT\right)^2}{2\sigma^2 T}$$

belongs to $\mathcal{M}_1$.
The $L^\infty(\mathbb{P}^{\text{hist}})$-condition on the payoff is needed to have

$$\sup_{\mathbb{Q} \in \mathcal{M}_1} \mathbb{E}^{\mathbb{Q}}[e^{-rT} |F_T|] < \infty$$

This could be replaced for example by $|F_T(S_T)| \leq K(1+S_T)$ for some constant K.
Theorem 1.1 means that the super-replication price can be obtained by maximizing the expectation of the discounted payoff (see the factor e^{-rT} multiplying F_T) over the (convex) set of all probabilities $\mathbb{Q}$ equivalent to $\mathbb{P}^{\text{hist}}$ and satisfying the constraint $\mathbb{E}^{\mathbb{Q}}[e^{-rT} S_T] = S_0$. Elements of $\mathcal{M}_1$ are called "risk-neutral" probabilities and are the dual variables associated to the delta-hedge H - see the proof below for an understanding of this terminology. In comparison to the historical/physical probability $\mathbb{P}^{\text{hist}}$ introduced to model the uncertainty of the asset price S_T, a risk-neutral probability $\mathbb{Q}$ should only be seen as a mathematical device. In particular, we should emphasize that in full generality (see e.g. Equation (1.4) where $\mathbb{P}^{\text{hist}}$ is a log-normal density with drift)

$$\mathbb{E}^{\mathbb{P}^{\text{hist}}}[e^{-rT} S_T] \neq S_0$$

In the discrete-time multi-period setting explained in the next section, it turns out that $e^{-rt} S_t$ will be a discrete $\mathbb{Q}$-martingale:

$$\mathbb{E}^{\mathbb{Q}}[e^{-rt_i} S_{t_i} | S_0, S_{t_1}, \ldots S_{t_{i-1}}] = e^{-rt_{i-1}} S_{t_{i-1}}$$

PROOF (Sketch) Here we give a formal proof as we take for granted a minimax argument (see below). For readers not familiar with the minimax principle, we think it is more important in first reading to understand the formal proof than to go along the lines of a rigorous proof which involves the Fenchel-Rockafeller theorem (Farkas's lemma for finite-dimensional linear program). Details can be found in [74] or [59] (see Theorem 2.4.1). A similar proof will be performed when discussing the Monge–Kantorovich duality in (M)OT. The derivation is done in three steps:

(i): In the first step, we introduce a positive measure $q(ds)$ having the same negligible sets as $\mathbb{P}^{\text{hist}}$. q is the Kuhn–Tucker multiplier associated to the inequality $\pi_T \geq 0$ $\mathbb{P}^{\text{hist}}$-almost surely. We write C_{sel} as the relaxed problem:

$$C_{\text{sel}} = \inf_{C,H} \sup_{q(\cdot) \geq 0} C - \int_0^\infty q(ds)\left(H\left(se^{-rT} - S_0\right) + C - e^{-rT}F(s)\right) \quad (1.9)$$

Indeed, note that if at some point s^* in the support of $\mathbb{P}^{\text{hist}}$, we have

$$H\left(s^* e^{-rT} - S_0\right) + C - e^{-rT}F(s^*) < 0$$

then the maximization over q is $+\infty$ by taking for q a Dirac measure supported on s^*: $q^*(ds) = \lambda\delta(s - s^*)ds$ with $\lambda \to \infty$. This is unsatisfactory as we must take an infimum (see the term $\inf_{C,H}$) in a second step. On the other hand, if $\pi_T \geq 0$ $\mathbb{P}^{\text{hist}}$-almost surely then $\pi_T \geq 0$ for all measures $q(ds)$ equivalent to $\mathbb{P}^{\text{hist}}$ and the maximum over q is zero. We have therefore proved that the relaxed version (1.9) of C_{sel} is identical to (1.7).

(ii): In the second step, we use a minimax argument which consists in switching the inf and sup operator (see Remark 1.2). We obtain

$$C_{\text{sel}} = \sup_{q(\cdot) \geq 0} \inf_{C,H} C + \int_0^\infty q(ds)\left(-H\left(se^{-rT} - S_0\right) - C + e^{-rT}F(s)\right)$$

(iii): By taking the infimum over C, we get $\int_0^\infty q(ds) = 1$, i.e., $\mathbb{Q} \equiv q(ds)$ is a probability measure equivalent to $\mathbb{P}^{\text{hist}}$: $\mathbb{Q} \sim \mathbb{P}^{\text{hist}}$. By taking the infimum over H, we get

$$\int_0^\infty q(ds)\left(se^{-rT} - S_0\right) = \mathbb{E}^{\mathbb{Q}}[(S_T e^{-rT} - S_0)] = 0,$$

i.e., $\mathbb{E}^{\mathbb{Q}}[e^{-rT}S_T] = S_0$. Finally, we obtain our dual formulation (1.8). ▯

REMARK 1.2 Weak/Strong duality   Remember that we always have the inequality

$$\inf_{x \in X} \sup_{y \in Y} f(x,y) \geq \sup_{y \in Y} \inf_{x \in X} f(x,y) \quad (1.10)$$

From our previous proof (see step **(ii)**), this implies that the *weak* duality result holds

$$C_{\text{sel}} \geq \sup_{\mathbb{Q} \in \mathcal{M}_1} \mathbb{E}^{\mathbb{Q}}[e^{-rT}F_T]$$

An alternative derivation goes as follows: By taking the expectation of the inequality $\pi_T \geq 0$, $\mathbb{P}^{\text{hist}}$ − a.s. with respect to $\mathbb{Q} \in \mathcal{M}_1$, we get

$$C \geq \mathbb{E}^{\mathbb{Q}}[e^{-rT} F_T]$$

We conclude by taking the supremum over $\mathbb{Q} \in \mathcal{M}_1$ and infimum over C. The *strong* duality stipulates that we have an equality. This is valid if for example: (1) f is continuous, concave (resp. convex) with respect to y for each x (resp. x for each y), (2) X, Y are compact convex sets (see [30] Chapter 1 for details with some weaker topological restrictions on X and Y). Note that when X, Y are finite-dimensional vector spaces and f is linear, the strong duality always hold. In our case, f is linear in $x = (C, H)$ and in $y = q(ds)$. By localization, X can be restricted to be a compact domain of $\mathbb{R}^2$. ⬜

Similarly, we can define *buyer's super-replication price* (also denoted the sub-replication price) as

DEFINITION 1.3 Buyer's super-replication price

$$C_{\text{buy}} \equiv \sup\{C \ : \ \exists\, H \text{ s.t. } - \pi_T \geq 0, \quad \mathbb{P}^{\text{hist}} - \text{a.s.}\} \tag{1.11}$$

We obtain the dual result

$$C_{\text{buy}} = \inf_{\mathbb{Q} \in \mathcal{M}_1} \mathbb{E}^{\mathbb{Q}}[e^{-rT} F_T] \tag{1.12}$$

1.2.2 Arbitrage-free prices and bounds

DEFINITION 1.4 Arbitrage opportunity $\mathbb{P}^{\text{hist}}$ *is arbitrage-free if there does not exist H and $\pi_0 < 0$ for which*

$$\pi_0 + H \left(S_T e^{-rT} - S_0\right) \geq 0, \quad \mathbb{P}^{\text{hist}} - \text{a.s.}$$

This means that starting with a strictly negative portfolio value at $t = 0$, a positive profit can be locked in without any downside risk. We say that we have an arbitrage opportunity. It is clear that from a modeling point of view, such arbitrage opportunity should be disregarded. Indeed, if such an opportunity would show up, it would generate a large demand on the underlying S, and at the equilibrium the arbitrage would disappear. Similarly, we define arbitrage-free prices as

DEFINITION 1.5 Arbitrage-free prices *We say that C is an arbitrage-free price if there does not exist H and $\pi_0 < 0$ for which*

$$\pi_0 + H \left(S_T e^{-rT} - S_0\right) + C - e^{-rT} F_T \geq 0, \quad \mathbb{P}^{\text{hist}} - \text{a.s.}$$

In fact these two definitions are consistent if we consider an extended market with two assets S_T and F_T and prices S_0 and C at $t = 0$. Throughout this book, we will assume

Assumption 2 $\mathbb{P}^{\mathrm{hist}}$ *is arbitrage-free.*

We have then the following lemma (easy to prove):

LEMMA 1.1
$\mathbb{P}^{\mathrm{hist}}$ *is not arbitrage-free if and only if $\hat{D} < 0$ with*

$$\hat{D} \equiv \inf\{\pi_0 \; : \; \exists \, H \text{ s.t. } \pi_0 + H\left(S_T e^{-rT} - S_0\right) \geq 0, \quad \mathbb{P}^{\mathrm{hist}} - \text{a.s.}\} \quad (1.13)$$

This corresponds to our previous linear programming problem (1.7) without holding an option (with payoff $F_T = 0$) but only performing a delta-hedging at $t = 0$. Playing the same game as before, this expression can be dualized (take $F_T = 0$ in Theorem 1.1) and we obtain $\hat{D} = \sup_{\mathbb{Q} \in \mathcal{M}_1} 0$ from which we imply:

COROLLARY 1.1
There is no arbitrage opportunity if and only if $\mathcal{M}_1 \neq \emptyset$.

REMARK 1.3 Duality and feasible convex set   Linear duality is a key tool for checking that a convex set $\mathcal{M}$ is non-empty - we say feasible. Indeed, we consider the linear programming problem $\sup_{x \in \mathcal{M}} 0$ that we dualize. Remember that by definition the supremum over an empty set is $-\infty$. If the dual problem is $-\infty$ then we conclude that $\mathcal{M} = \emptyset$.   ⬜

THEOREM 1.2
Let C be an arbitrage-free price. Then,
(i):

$$C_{\mathrm{buy}} = \inf_{\mathbb{Q} \in \mathcal{M}_1} \mathbb{E}^{\mathbb{Q}}[e^{-rT} F_T] \leq C \leq C_{\mathrm{sel}} = \sup_{\mathbb{Q} \in \mathcal{M}_1} \mathbb{E}^{\mathbb{Q}}[e^{-rT} F_T]$$

(ii): *There exists $\mathbb{Q}^* \in \overline{\mathcal{M}}_1$ (the closure of $\mathcal{M}_1$) such that*

$$C = \mathbb{E}^{\mathbb{Q}^*}[e^{-rT} F_T]$$

PROOF   **(i)**: By contradiction, let us assume $C > C_{\mathrm{sel}}$. We lock up into an arbitrage by selling the option at the price C and by entering into the super-replication strategy at the price C_{sel}. Similar argument for proving that $C \geq C_{\mathrm{buy}}$. **(ii)**: As $\mathcal{M}_1$ is relatively compact (from Prokhorov's theorem - See Remark 1.4), C_{buy} and C_{sel} are attained respectively by $\mathbb{Q}^{\mathrm{buy}}$ and $\mathbb{Q}^{\mathrm{sel}}$

in $\overline{\mathcal{M}}_1$. By convex combination $\mathbb{Q}^\theta \equiv \theta\mathbb{Q}^{\text{buy}} + (1-\theta)\mathbb{Q}^{\text{sel}}$ for $\theta \in [0,1]$, we can reach any $C \in [C_{\text{buy}}, C_{\text{sel}}]$. □

REMARK 1.4 Prokhorov's theorem states that any tight family $\mathcal{M}$ in $\mathcal{P}(\mathbb{R}^n)$ (the set of probability measures on $\mathbb{R}^n$) is relatively compact in $\mathcal{P}(\mathbb{R}^n)$: from every sequence $\mathbb{Q}_k \in \mathcal{P}$ one can extract a subsequence $\mathbb{Q}_{n_k} \rightharpoonup \mathbb{Q}$ (i.e., for any $\psi \in C_b$, $\lim_{k\to\infty} \mathbb{E}^{\mathbb{Q}_{n_k}}[\psi] = \mathbb{E}^{\mathbb{Q}}[\psi]$). A family $\mathcal{M}$ of $\mathcal{P}$ is said to be tight if for any $\epsilon > 0$, there exists a compact subset K_ϵ for which $\sup_{\mathbb{Q}\in\mathcal{M}} \mathbb{Q}(\mathbb{R}^n \setminus K_\epsilon) \leq \epsilon$.

As for all $\mathbb{Q} \in \mathcal{M}_1$, $\mathbb{Q}[S_T \geq \epsilon] \leq \mathbb{E}[S_T]/\epsilon = S_0 e^{rT}/\epsilon$, we conclude the relatively compactness of $\mathcal{M}_1$. In particular for $\mathbb{Q} \in \overline{\mathcal{M}}_1$, $\mathbb{E}^{\mathbb{Q}}[e^{-rT}S_T] = S_0$ but in general $\mathbb{Q}$ is not equivalent to $\mathbb{P}^{\text{hist}}$. □

REMARK 1.5 $\mathbb{Q}^{\text{sel}}$ (resp. $\mathbb{Q}^{\text{buy}}$) depends on F_T and therefore $\mathbb{E}^{\mathbb{Q}^{\text{buy}}}[e^{-rT} \cdot]$ is not necessary a linear operator on the set of payoff functions! The buyer's super-replication prices of F_T^1 and F_T^2 is not equal to the super-replication price of $F_T^1 + F_T^2$:

$$\mathbb{E}^{\mathbb{Q}^{\text{buy}}}[e^{-rT}(F_T^1 + F_T^2)] \neq \mathbb{E}^{\mathbb{Q}^{\text{buy}}}[e^{-rT}F_T^1] + \mathbb{E}^{\mathbb{Q}^{\text{buy}}}[e^{-rT}F_T^2]$$

□

COROLLARY 1.2

The arbitrage-free price of F_T is unique if and only if $\mathcal{M}_1$ reduces to a singleton.

1.2.3 A worked-out example: The binomial model

Let us assume that the historical probability $\mathbb{P}^{\text{hist}}$ is supported on two points $S_u = uS_0$ and $S_d = dS_0$ with probabilities p_u and p_d and $u > d$ without loss of generality. Note that $e^{-rT}S_T$ is not required to be a $\mathbb{P}^{\text{hist}}$-martingale. The price obtained using an insurance point of view is (see Formula (1.2))

$$C_{\text{ins}} = e^{-rT}\left(p_u(uS_0 - K)^+ + p_d(dS_0 - K)^+\right)$$

The super-replication price for a call can be obtained easily using our dual formulation. Let us characterize the convex set $\mathcal{M}_1$: $\mathbb{Q} \in \mathcal{M}_1$ if and only if $\mathbb{Q}$ is supported on the points S_u and S_d (as $\mathbb{Q} \sim \mathbb{P}^{\text{hist}}$) and satisfies $\mathbb{E}^{\mathbb{Q}}[e^{-rT}S_T] = S_0$:

$$q_u + q_d = 1$$
$$q_u S_u + q_d S_d = e^{rT}S_0$$

where $q_u \equiv \mathbb{Q}(S_T = S_u) \neq 0$ and $q_d \equiv \mathbb{Q}(S_T = S_d) \neq 0$. There is a (unique) solution if and only if $d < e^{rT} < u$ for which

$$q_u = \frac{e^{rT} - d}{u - d}, \quad q_d = \frac{u - e^{rT}}{u - d}$$

From Corollary 1.1, the binomial model is arbitrage-free if and only if $d < e^{rT} < u$. The super-replication price is then

$$C = e^{-rT} \left(q_u(uS_0 - K)^+ + q_d(dS_0 - K)^+ \right) \tag{1.14}$$

Note that as explained previously, C depends only on the negligible set of $\mathbb{P}^{\text{hist}}$ (through the points S_d and S_u) and does not depend on p_u and p_d. As the set $\mathcal{M}_1$ is a singleton, the price of this option is unique from Corollary 1.2 (in particular the super and sub-replication prices coincide).

PROPOSITION 1.1 Perfect replication

$$e^{-rT} \pi_T \equiv C + H^*(S_T e^{-rT} - S_0) - e^{-rT}(S_T - K)^+ = 0, \quad \mathbb{P}^{\text{hist}} - \text{a.s.}$$

with C given by (1.14) and

$$H^* = \frac{(uS_0 - K)^+ - (dS_0 - K)^+}{(u - d)S_0}$$

PROOF As C is both a super-replication price and a sub-replication price, we have $\pi_T \geq 0$ and $\pi_T \leq 0$ $\mathbb{P}^{\text{hist}} - \text{a.s.}$. In particular, $\pi_T = 0$ on the events $S_T = S_d$ and $S_T = S_u$. This gives the expression for C and H. ▢

Proposition 1.1 has a clear interpretation from a trading point of view. It means that if $\mathbb{P}^{\text{hist}}$ is (properly) described by a binomial model, the payoff F_T can be perfectly replicated by holding at $t = 0$ H^* units of the share. This is a general result that we will prove below: if $\mathcal{M}_1$ reduces to a singleton, the option can be perfectly replicated by performing a delta-hedging strategy.

1.2.4 Replication paradigm

DEFINITION 1.6 Attainable payoff *A payoff F_T is attainable/replicable if there exists C and H such that*

$$e^{-rT} \pi_T \equiv C + H(S_T e^{-rT} - S_0) - e^{-rT} F_T = 0, \quad \mathbb{P}^{\text{hist}} - \text{a.s.}$$

We say that the payoff F_T is perfectly replicated by using a delta-hedging strategy. Here the trader faces no risk, modulo the negligible sets of $\mathbb{P}^{\text{hist}}$ are properly modeled.

DEFINITION 1.7 Complete model $\mathbb{P}^{\text{hist}}$ *is called a complete model if every payoff is attainable.*

THEOREM 1.3
$\mathbb{P}^{\text{hist}}$ *is a complete model if and only if* $\mathcal{M}_1$ *reduces to a singleton.*

PROOF
$\Longrightarrow$ By definition, we have for a payoff F_T, that there exists C and H such that

$$C + H(S_T e^{-rT} - S_0) - e^{-rT} F_T = 0, \quad \mathbb{P}^{\text{hist}} - \text{a.s.}$$

By taking the expectation with respect to $\mathbb{Q} \in \mathcal{M}_1$, this implies that $C = \mathbb{E}^{\mathbb{Q}}[e^{-rT} F_T]$. In particular for $\mathbb{Q}_1, \mathbb{Q}_2 \in \mathcal{M}_1$, we have $\mathbb{E}^{\mathbb{Q}_1}[F_T] = \mathbb{E}^{\mathbb{Q}_2}[F_T]$. This implies that $\mathbb{Q}_1 = \mathbb{Q}_2$ from the arbitrariness of F_T.
$\Longleftarrow$ Let $\mathcal{M}_1 = \{\mathbb{Q}^*\}$. From Corollary 1.2, the arbitrage-free price of a payoff F_T, denoted C, is unique and we have a sub and super-replication strategy:

$$e^{-rT} \pi_T^{\text{super}} \equiv C + H_{\text{super}}(S_T e^{-rT} - S_0) - e^{-rT} F_T \geq 0, \quad \mathbb{Q}^* - \text{a.s.}$$
$$e^{-rT} \pi_T^{\text{sub}} \equiv C + H_{\text{sub}}(S_T e^{-rT} - S_0) - e^{-rT} F_T \leq 0, \quad \mathbb{Q}^* - \text{a.s.}$$

for some delta H_{super} and H_{sub}. By taking the expectation with respect to $\mathbb{Q}^*$, we get that π_T^{super} and π_T^{sub} have zero mean and therefore the above inequalities are equalities. As $\mathbb{Q}^* \sim \mathbb{P}^{\text{hist}}$, we get that F_T is attainable by (C, H_{super}) (and also by (C, H_{sub})) from which we can conclude. Note that by subtracting π_T^{super} and π_T^{super}, we get

$$(H_{\text{super}} - H_{\text{sub}}) \left(S_T e^{-rT} - S_0 \right) = 0, \quad \mathbb{Q}^* - \text{a.s.}$$

This implies that $H_{\text{super}} = H_{\text{sub}}$. $\quad\square$

Theorems 1.2 and Corollary 1.2, answer our first question (A). Theorem 1.3 answers the second question (B) in the case of a complete model.

1.2.5 Geometry of $\mathcal{M}_1$: Extremal points

$\mathcal{M}_1$ is a convex set. A point $\mathbb{Q} \in \mathcal{M}_1$ is said to be extremal if it cannot be decomposed as a convex combination of two elements in $\mathcal{M}_1$: for $\theta \in (0, 1)$, $\mathbb{Q} = \theta \mathbb{Q}^1 + (1-\theta) \mathbb{Q}^2 \Longrightarrow \mathbb{Q}^1 = \mathbb{Q}^2$. As $\mathcal{M}_1$ is relatively compact, one can show (see Minkowski or Krein–Milman theorems) that $\overline{\mathcal{M}}_1$ can be reconstructed as the convex hull of its extremal points. Furthermore, we cite the technical lemma (easy to understand - take x^2 on $[0, 1]$; see Lemma 4.1.12 in [27] for a proof)

LEMMA 1.2 Bauer maximum principle
Let f be a convex continuous function over a compact convex set K. It follows that f attains its maximum at an extremal point of K.

This means that when maximizing the (linear) functional $\mathbb{E}^{\mathbb{Q}}[e^{-rT}F_T]$ over $\mathbb{Q} \in \mathcal{M}_1$, the supremum is therefore attained by an extremal point $\mathbb{Q}^*$ in $\overline{\mathcal{M}}_1$. The payoff F_T is then attainable for $\mathbb{Q}^*$. Indeed $e^{-rT}\pi_T$ is a positive random variable as being a super-replication strategy and has a zero $\mathbb{Q}^*$-mean (as $\mathbb{E}^{\mathbb{Q}^*}[e^{-rT}\pi_T] = C - \mathbb{E}^{\mathbb{Q}^*}[e^{-rT}F_T] = 0$). Therefore, $\pi_T = 0$ $\mathbb{Q}^*$-a.s.

At this point, the main question left is

(C) If $\mathbb{P}^{\text{hist}}$ is not a complete model, how can we choose an element $\mathbb{Q}^* \in \overline{\mathcal{M}}_1$?

In the next sections, we will select a (decent) $\mathbb{Q}^* \in \overline{\mathcal{M}}_1$ by looking at some particular utility functions. In fact, all risk-neutral measures can be constructed by choosing an appropriate utility function (in the dual formulation) - see Lemma 1.2.

1.2.6   Mean-variance: Quadratic programming

By choosing a log-normal distribution (with drift μ and volatility σ) for $\mathbb{P}^{\text{hist}}$, we can show that the seller's super-replication price of a call option is S_0. The reader should remark that from Jensen's inequality we have also for all $\mathbb{Q} \in \mathcal{M}_1$,

$$C_{\text{buy}} = (S_0 - Ke^{-rT})^+ \leq \mathbb{E}^{\mathbb{Q}}[e^{-rT}(S_T - K)^+] \leq C_{\text{sel}} = S_0 \qquad (1.15)$$

This seller super-replication's price is very expensive as it is identical to the price of a forward contract that it is the option that delivers S_T at the maturity. It is therefore fairly unexpected that a (reasonable) client is willing to accept to pay a call option at the same price as a forward contract for which the payoff super-replicates at maturity those of a call option $(S_T - K)^+ \leq S_T$. In this section, we disregard the super-replication approach in this respect and fix C and H such that the variance of π_T for a payoff F_T is minimised under the constraint $\mathbb{E}^{\mathbb{P}^{\text{hist}}}[\pi_T] = 0$ (i.e., fixed return):

DEFINITION 1.8 Mean-variance hedging   *The mean-variance hedging is defined as the following quadratic programming problem:*

$$P_{\text{quad}} \equiv \inf_{C, H \text{ s.t. } \mathbb{E}^{\mathbb{P}^{\text{hist}}}[\pi_T]=0} \mathbb{E}^{\mathbb{P}^{\text{hist}}}[\pi_T^2] \qquad (1.16)$$

As the cost $\mathbb{E}^{\mathbb{P}^{\text{hist}}}[\pi_T^2]$ (resp. the constraint $\mathbb{E}^{\mathbb{P}^{\text{hist}}}[\pi_T] = 0$) is a quadratic (resp. linear) form with respect to C and H, (1.16) defines a *quadratic programming problem.* The constraint $\mathbb{E}^{\mathbb{P}^{\text{hist}}}[\pi_T] = 0$ gives

$$C_{\text{quad}} = e^{-rT}\mathbb{E}^{\mathbb{P}^{\text{hist}}}[F_T] - H\mathbb{E}^{\mathbb{P}^{\text{hist}}}[(S_T e^{-rT} - S_0)]$$

where C_{quad} is the minimizer in (1.16). The first term corresponds to our insurance price C_{ins}. Computing the infimum over H of $\mathbb{E}[\pi_T^2]$ with $C = C_{\text{quad}}$, we obtain

$$\mathbb{E}^{\mathbb{P}^{\text{hist}}}\left[\pi_T\left(S_T - \mathbb{E}^{\mathbb{P}^{\text{hist}}}[S_T]\right)\right] = 0$$

where we have used $e^{-rT}\partial_H \pi_T = e^{-rT}\left(S_T - \mathbb{E}^{\mathbb{P}^{\text{hist}}}[S_T]\right)$. This is equivalent to

$$H_{\text{quad}} = \frac{\mathbb{E}^{\mathbb{P}^{\text{hist}}}[F_T(S_T - \mathbb{E}^{\mathbb{P}^{\text{hist}}}[S_T])]}{\mathbb{E}^{\mathbb{P}^{\text{hist}}}[(S_T - \mathbb{E}^{\mathbb{P}^{\text{hist}}}[S_T])(S_T - S_0 e^{rT})]} \tag{1.17}$$

Finally,

PROPOSITION 1.2

P_{quad} *is attained by the (arbitrage-free) price*

$$C_{\text{quad}} = e^{-rT}\mathbb{E}^{\mathbb{Q}^{\text{quad}}}[F_T]$$

where

$$\frac{d\mathbb{Q}^{\text{quad}}}{d\mathbb{P}^{\text{hist}}} = \left(1 - \frac{(S_T - \mathbb{E}^{\mathbb{P}^{\text{hist}}}[S_T])\mathbb{E}^{\mathbb{P}^{\text{hist}}}[(S_T - S_0 e^{rT})]}{\mathbb{E}^{\mathbb{P}^{\text{hist}}}[(S_T - \mathbb{E}^{\mathbb{P}^{\text{hist}}}[S_T])(S_T - S_0 e^{rT})]}\right)$$

and H_{quad} given by (1.17).

We can check that $\mathbb{Q}^{\text{quad}} \in \mathcal{M}_1$. Note that from the inequality (1.15), we have for a call option:

$$(S_0 - Ke^{-rT})^+ \leq C_{\text{quad}} \leq S_0$$

As a numerical application, for $r = 5\%$, $\mu = 10\%$, $\sigma = 20\%$ and $S_0 = 100$, the price of a call option with $T = 1$ year and $K/S_0 = 1$ from Proposition 1.2 is $C_{\text{quad}} = 10.06$ ($H_{\text{quad}} = 0.76$). The residual standard deviation is $\sqrt{\mathbb{E}^{\mathbb{P}^{\text{hist}}}[\pi_T^2]} = 5.53$.

An alternative definition of the mean-variance hedging could be

DEFINITION 1.9 Mean-variance hedging, again *Fix $\zeta \in \mathbb{R}$.*

$$P_{\text{quad}}(\zeta) \equiv \inf_{C, H}\left(\mathbb{E}^{\mathbb{P}^{\text{hist}}}[\pi_T^2] - \mathbb{E}^{\mathbb{P}^{\text{hist}}}[\pi_T]^2\right) + \zeta\mathbb{E}^{\mathbb{P}^{\text{hist}}}[\pi_T]^2 \tag{1.18}$$

Trivially, $P_{\text{quad}}(\infty) = P_{\text{quad}}$.

1.2.7 Utility function: Convex programming

The above approach can be generalized by considering

$$P_{\text{concave}} = \sup_{C, H} \mathbb{E}^{\mathbb{P}^{\text{hist}}}[U(\pi_T)] \tag{1.19}$$

where U is concave and negative. Note that the sub-replication approach can be obtained by taking $U(x) = -\infty$ if $x > 0$ and $U(x) = 0$ otherwise. This enforces the condition $-\pi_T \geq 0$, $\quad \mathbb{P}^{\text{hist}} - \text{a.s.}$.

1.2.8 Quantile hedging

Quantile hedging consists in replacing Definition 1.1 of the seller's price by the following:

$$C_p = \inf\{C \; : \; \exists \, H \text{ s.t. } \mathbb{P}^{\text{hist}}[\pi_T \geq 0] \geq p$$
$$\text{and } C + H(S_T e^{-rT} - S_0) \geq 0, \quad \mathbb{P}^{\text{hist}} - \text{a.s.}\}$$

$p \in [0,1]$ is interpreted as the probability of super-replicating the claim F_T under the historical measure. Here we have added the constraint that the trader's portfolio should be greater than a threshold $-L$:

$$C + H(S_T e^{-rT} - S_0) \geq -L$$

For convenience, we have chosen $L = 0$, this can be easily modified.

By definition, $C_1 = C_{\text{sel}}$, the super-replication price for $F_T \geq 0$. C_{sel} can be very high – recall for instance that the super-replication price of a call option equals S_0. In the quantile hedging approach, we only impose that the payoff can be super-replicated with a probability p. In their seminal paper [75], Föllmer and Leukert show that the corresponding optimal strategy consists in superhedging a modified payoff. More precisely, we have

THEOREM 1.4 Duality - quantile hedging

$$C_p = \inf_{A \in \mathcal{F}_T \text{ s.t. } \mathbb{P}^{\text{hist}}[A] \geq p} \; \sup_{\mathbb{Q} \in \mathcal{M}_1} \mathbb{E}^{\mathbb{Q}}[e^{-rT} F_T 1_A]$$

PROOF (Sketch) C_p can be written as a classical super-replication problem:

$$C_p = \inf_{A \in \mathcal{F}_T \text{ s.t. } \mathbb{P}^{\text{hist}}[A] \geq p} \inf\{C \; : \; \exists \, H \text{ s.t. } C + H(S_T e^{-rT} - S_0) \geq e^{-rT} F_T 1_A,$$
$$\mathbb{P}^{\text{hist}} - \text{a.s.}\}$$

This means that we super-replicate the payoff F_T in A, outside $C + H(S_T e^{-rT} - S_0) \geq 0$. The set A is such that $\mathbb{P}^{\text{hist}}[A] \geq p$. This problem can then be dualized as in Theorem 1.1 and we get our result. $\qquad \square$

1.2.9 Utility indifference price

We introduce an utility function U, which is strictly increasing and concave. We consider the value $u(x,0)$ of the supremum over all hedging portfolios starting from the initial capital x of the expectation of the utility of the discounted final wealth under the historical measure $\mathbb{P}^{\text{hist}}$:

$$u(x,0) \equiv \sup_H \mathbb{E}^{\mathbb{P}^{\text{hist}}} \left[U\left(x - H(e^{-rT}S_T - S_0) \right) \right]$$

Similarly, the value $u(x - C, F_T)$ is defined for a claim F_T as

$$u(x - C, F_T) \equiv \sup_H \mathbb{E}^{\mathbb{P}^{\text{hist}}} \left[U\left(x - C + e^{-rT}F_T - H(e^{-rT}S_T - S_0) \right) \right] \quad (1.20)$$

The utility indifference buyer's price, as introduced by Hodges-Neuberger [107], is the quantity C_{HN} such that

DEFINITION 1.10 Utility indifference buyer's price

$$u(x,0) = u(x - C_{\text{HN}}, F_T)$$

This means that a buyer should accept quoting a price for the claim F_T when buying and delta-hedging this derivative becomes as profitable as setting up a pure delta strategy. The expression $u(x - C, F_T)$ can be dualized into

THEOREM 1.5

$$u(x - C, F_T) = \inf_{\mathbb{Q} \in \mathcal{M}_1} \mathbb{E}^{\mathbb{Q}} \left[(e^{-rT}F_T + x - C) + \frac{d\mathbb{P}^{\text{hist}}}{d\mathbb{Q}} U^* \left(\frac{d\mathbb{Q}}{d\mathbb{P}^{\text{hist}}} \right) \right]$$

$$(1.21)$$

with $U^*(p) \equiv \sup_{x \in \mathbb{R}}\{U(x) - px\}$ the Legendre-Fenchel transform of U. The functions U and U^* also satisfy the conjugate relation: $U(x) = \inf_{p \in \mathbb{R}_+}\{px + U^*(p)\}$.

PROOF (Sketch) By introducing the Lagrange multiplier $H(S_T e^{-rT} - S_0)$, the infimum over $\mathbb{Q} \in \mathcal{M}_1$ in Equation (1.21) can be relaxed as an infimum over the space of probabilities (equivalent to $\mathbb{P}^{\text{hist}}$):

$$I \equiv \inf_{\mathbb{Q} \sim \mathbb{P}^{\text{hist}}} \sup_H \mathbb{E}^{\mathbb{P}^{\text{hist}}} \left[(-e^{-rT}\pi_T + x) \frac{d\mathbb{Q}}{d\mathbb{P}^{\text{hist}}} + U^* \left(\frac{d\mathbb{Q}}{d\mathbb{P}^{\text{hist}}} \right) \right]$$

We can then invoke a minimax argument (see Theorem 1.1 for a similar proof) and I can be written as

$$I = \sup_H \inf_{\mathbb{Q} \sim \mathbb{P}^{\text{hist}}} \mathbb{E}^{\mathbb{P}^{\text{hist}}} \left[(-e^{-rT}\pi_T + x) \frac{d\mathbb{Q}}{d\mathbb{P}^{\text{hist}}} + U^* \left(\frac{d\mathbb{Q}}{d\mathbb{P}^{\text{hist}}} \right) \right]$$

By taking the infimum over $\mathbb{Q} \sim \mathbb{P}^{\text{hist}}$ (and using the definition of U^*), we reproduce our definition (1.20). □

Example 1.1 Exponential utility, $\alpha > 0$
Choose

$$U(x) = 1 - e^{-\alpha x}, \quad U^*(p) = 1 + \frac{p}{\alpha}\left(\ln\frac{p}{\alpha} - 1\right)$$

Then, from Equation (1.21), the indifference price, which does not depend on the initial portfolio value x, is

$$\alpha C_{\text{HN}}^\alpha = \inf_{\mathbb{Q} \in \mathcal{M}_1}\left\{\alpha \mathbb{E}^\mathbb{Q}[e^{-rT}F_T] + H(\mathbb{Q}, \mathbb{P}^{\text{hist}})\right\} - \inf_{\mathbb{Q} \in \mathcal{M}_1} H(\mathbb{Q}, \mathbb{P}^{\text{hist}})$$

where $H(\mathbb{Q}, \mathbb{P}^{\text{hist}}) \equiv \mathbb{E}^\mathbb{Q}[\ln\frac{d\mathbb{Q}}{d\mathbb{P}^{\text{hist}}}]$ is the relative entropy of $\mathbb{Q}$ with respect to $\mathbb{P}^{\text{hist}}$. Note that

$$\lim_{\alpha \to \infty} C_{\text{HN}}^\alpha = C_{\text{buy}}$$

and

$$\lim_{\alpha \to 0} C_{\text{HN}}^\alpha = \mathbb{E}^{\mathbb{Q}^{\text{ent}}}[e^{-rT}F_T]$$

where $\mathbb{Q}^{\text{ent}}$ is among all probabilities in $\mathcal{M}_1$ that one which minimises the relative entropy:

$$\mathbb{Q}^{\text{ent}} = \text{argmin}_{\mathbb{Q} \in \mathcal{M}_1} H(\mathbb{Q}, \mathbb{P}^{\text{hist}})$$

$\mathbb{Q}^{\text{ent}}$ is called the minimal martingale entropy probability and is explicitly given by

$$\frac{d\mathbb{Q}^{\text{ent}}}{d\mathbb{P}^{\text{hist}}} = \frac{e^{-H^*(S_T e^{-rT} - S_0)}}{\mathbb{E}^{\mathbb{P}^{\text{hist}}}[e^{-H^*(S_T e^{-rT} - S_0)}]}$$

with H^* the unique solution of $H^* = \text{argmax}_H - \ln \mathbb{E}^{\mathbb{P}^{\text{hist}}}[e^{-H(S_T e^{-rT} - S_0)}]$. $\mathbb{Q}^{\text{ent}}$ is an example of Gibbs measure. We can check that $\mathbb{Q}^{\text{ent}} \in \mathcal{M}_1$. □

1.2.10 A worked-out example: The trinomial model

Let us consider a trinomial model: $S_0 = 1$, and S_T can take only three values at maturity T: 0.5, 1, and 1.5. Under the historical measure, we have $\mathbb{P}^{\text{hist}}(S_T = 0.5) = 1/3$, $\mathbb{P}^{\text{hist}}(S_T = 1.0) = 1/3$, and $\mathbb{P}^{\text{hist}}(S_T = 1.5) = 1/3$. For the sake of simplicity, we assume here a zero interest rate. $\mathbb{Q} \in \mathcal{M}_1$ if and only if $q_{-1} \equiv \mathbb{Q}(S_T = 0.5)$, $q_0 \equiv \mathbb{Q}(S_T = 1)$ and $q_1 \equiv \mathbb{Q}(S_T = 1.5)$ satisfy the 2 linear equations:

$$q_{-1} + q_0 + q_1 = 1, 0.5q_{-1} + q_0 + 1.5q_1 = 1$$

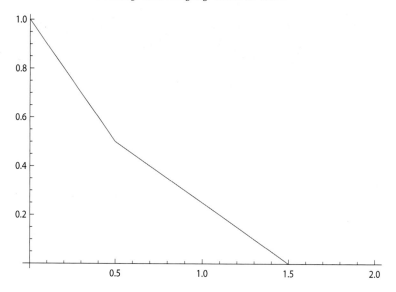

Figure 1.2: $C_{\mathrm{sel}}(K)$ as a function of K for the trinomial model with $\mathbb{P}^{\mathrm{hist}}(S_T = 0.5) = \mathbb{P}^{\mathrm{hist}}(S_T = 1.0) = \mathbb{P}^{\mathrm{hist}}(S_T = 1.5) = 1/3$ and $r = 0$.

This implies that

$$\mathcal{M}_1 = \{(q_{-1}, q_0, q_1) \; : \; q_{-1} = q_1 \in (0, 1/2), q_0 = 1 - 2q_1\}$$

The extremal points of $\overline{\mathcal{M}}_1$ are $(1/2, 0, 1/2)$ which corresponds to the binomial model and $(0, 1, 0)$.

(i): The super-replication price of a call option with maturity T and strike K is

$$C_{\mathrm{sel}}(K) \equiv \inf\{C \; : \; \exists\, H \text{ s.t. } C + H(S_T - S_0) \geq (S_T - K)^+, \quad \mathbb{P}^{\mathrm{hist}} - \text{a.s}\}$$

This is equivalent to the (finite-dimensional) linear program:

$$C_{\mathrm{sel}}(K) \equiv \inf_C C$$

such that $C - 0.5H \geq (0.5 - K)^+$, $C \geq (1 - K)^+$ and $C + 0.5H \geq (1.5 - K)^+$. A numerical solution gives Figure 1.2.

(ii): The quantile hedging price as a function of $p \in [0, 1]$ can be written as

$$C_{\mathrm{quantile}}(p) \equiv \inf_{C, H, (e_i)_{i=1,2,3} \in \{0,1\}} C$$

such that $\sum_{i=1}^3 e_i p_i \geq p$ with $C \geq 0.5|H|$ and

$$e_1(C - 0.5H - (0.5 - K)^+) \geq 0, \quad e_2(C - (1 - K)^+) \geq 0$$
$$e_3(C + 0.5H - (1.5 - K)^+) \geq 0$$

The constraint $\mathbb{P}^{\text{hist}}[\pi_T \geq 0] \geq p$ has been linearized by introducing the boolean variables $(e_i)_{i=1,2,3} \in \{0,1\}$. For fixed $(e_i)_{i=1,2,3}$, this is a linear program that can be solved. By listing all the combinations for e_i (i.e., 2^3), $C_{\text{quantile}}(p)$ can be solved. For example, we get for $K = 1$:

$$C_{\text{quantile}}(1/3) = C_{\text{quantile}}(2/3) = 0, \quad C_{\text{quantile}}(1) = C_{\text{sel}}(1) = 0.25$$

(iii): The quadratic hedging price is

$$C_{\text{quad}}(K) \equiv \inf_{C,H} \mathbb{E}^{\mathbb{P}^{\text{hist}}}[(C + H(S_T - S_0) - (S_T - K)^+)^2]$$

such that $\mathbb{E}^{\mathbb{P}^{\text{hist}}}[(C + H(S_T - S_0) - (S_T - K)^+)] = 0$. We find $C_{\text{quad}}(1) = 1/6$ (for which $H = 1/2$).
As a conclusion, for $K = 1$, $C_{\text{quad}} = 1/6$, $C_{\text{quantile}}(1/3) = 0$, $C_{\text{quantile}}(2/3) = 0$ and $C_{\text{quantile}}(1) = C_{\text{sel}} = 1/4$.

1.3   A cautious trader viewpoint

Our delta-hedging strategy was a bit too simple as it consists only in buying or selling the asset at $t = 0$ and holding it until the maturity. A more involved strategy is to buy (or sell) units of the asset at a date t_k until a next date t_{k+1}. Let us compute the value of our delta-hedged portfolio at the maturity T. At t_k, the portfolio value π_{t_k} is

$$\pi_{t_k} = (\pi_{t_k} - H_{t_k} S_{t_k}) + H_{t_k} S_{t_k}$$

where H_{t_k} is the number of shares held at time t_k. Although this expression seems algebraically trivial, its financial interpretation is important: the term $H_{t_k} S_{t_k}$ is the value at t_k of a position consisting of H_{t_k} units of the asset. The term $\pi_{t_k} - H_{t_k} S_{t_k}$ represents the cash part invested in a bank account. The variation of our portfolio between t_k and t_{k+1} is then

$$\delta\pi_{t_k} = (\pi_{t_k} - H_{t_k} S_{t_k}) \, r\delta t + H_{t_k} \delta S_{t_k}$$
$$= \pi_{t_k} r\delta t + H_{t_k}(\delta S_{t_k} - S_{t_k} r\delta t)$$

with $\delta S_{t_k} \equiv S_{t_{k+1}} - S_{t_k}$, $\delta t = t_{k+1} - t_k$ small enough. As no cash is injected between t_k and t_{k+1}, our portfolio is called *self-financing*. By setting $\tilde{\pi}_{t_k} \equiv e^{-rt_k}\pi_{t_k}$ and $\tilde{S}_{t_k} \equiv e^{-rt_k}S_{t_k}$, we obtain the variation of the discounted portfolio

$$\delta\tilde{\pi}_{t_k} = H_{t_k}\delta\tilde{S}_{t_k}, \quad \delta\tilde{S}_{t_k} \equiv \tilde{S}_{t_{k+1}} - \tilde{S}_{t_k}$$

Here the state of information evolves over time and is described by a filtration $\mathcal{F} = (\mathcal{F}_{t_1}, \ldots, \mathcal{F}_{t_n})$ where the σ-algebra $\mathcal{F}_t$ is the set of events that will be known to be true or false. We take here $\mathcal{F}_{t_k} = \sigma(S_0, \ldots, S_{t_k})$ the natural filtration. $H_k = H_k(S_0, \ldots, S_{t_k})$ is adapted, i.e., a measurable function with respect to $\mathcal{F}_{t_k}$: we don't look into the future. If we now assume that the trader sells an option with payoff F_T at the price C at $t = 0$ and then delta-hedges his position at the intermediate dates $t_0 \equiv 0 < t_1 < \ldots < t_n \equiv T$, we get

$$e^{-rT} \pi_T = -e^{-rT} F_T + C + \sum_{k=0}^{n-1} H_{t_k}(S_0, \ldots, S_{t_k}) \delta \tilde{S}_{t_k} \qquad (1.22)$$

By playing the same game as in Theorem 1.1, we obtain the dual expression:

THEOREM 1.6
Let $F_T \in L^\infty(\mathbb{P}^{\text{hist}})$. Then,

$$C_{\text{sel}} = \sup_{\mathbb{Q} \in \mathcal{M}_n} \mathbb{E}^{\mathbb{P}}[e^{-rT} F_T] \qquad (1.23)$$

with

$$\mathcal{M}_n \equiv \{ \mathbb{Q} : \mathbb{Q} \sim \mathbb{P}^{\text{hist}}, \quad \mathbb{E}^{\mathbb{Q}}[e^{-rt_k} S_{t_k} | \mathcal{F}_{t_{k-1}}] = e^{-rt_{k-1}} S_{t_{k-1}},$$
$$\forall\, k = 1, \ldots, n \} \qquad (1.24)$$

An element $\mathbb{Q}$ of $\mathcal{M}_n$ is called an equivalent (to $\mathbb{P}^{\text{hist}}$) martingale measure. The condition $\mathbb{E}^{\mathbb{Q}}[e^{-rt_k} S_{t_k} | \mathcal{F}_{t_{k-1}}] = e^{-rt_{k-1}} S_{t_{k-1}}$ indicates that the process $e^{-rt} S_t$ is a (discrete) martingale under $\mathbb{Q}$.

REMARK 1.6 Markov property   If we impose that our delta-hedging H_{t_k} depends only on S_{t_k}, $H_{t_k} = H(S_{t_k})$ - say that the trader does not want to store the path history of the asset in order to compute his delta-hedging - then in Equation (1.24), $\mathcal{M}_n$ is replaced by the simpler condition:

$$\mathbb{E}^{\mathbb{Q}}[e^{-rt_k} S_{t_k} | S_{t_{k-1}}] = e^{-rt_{k-1}} S_{t_{k-1}}$$

This indicates that $e^{-rt} S_t$ is a discrete martingale with the Markov property.
⬚

As an exercise, we can re-derive results in the previous sections in the case of discrete delta-hedging (and also in the continuous-time setup by formally taking the limit $n \to \infty$). The proofs are highly technical in the continuous-time setup and require a whole book (see [59]). Theorems 1.1, 1.2, 1.3 and Corollary 1.2, remain valid in this setup if we replace $\mathcal{M}_1$ by $\mathcal{M}_n$ (or $\mathcal{M}_\infty$). We should remark that $\mathcal{M}_\infty \subset \mathcal{M}_n \subset \mathcal{M}_1$ and therefore a model $\mathbb{P}^{\text{hist}}$ which is not complete by assuming only delta-hedging at $t = 0$ can be completed if we assume discrete-time or even continuous-time delta-hedging. This is precisely the case for the Black–Scholes model which assumes that the asset price is modeled by a log-normal process and which is complete in the continuous-time delta-hedging. When $n \to \infty$, our hedge portfolio value (1.22) converges (in a certain sense - see a course in stochastic integration) to

$$e^{-rT}\pi_T = -e^{-rT}F_T + C + \int_0^T H_t d\tilde{S}_t$$

S_t is required to be a semi-martingale in order to get a well-defined (stochastic) integral $\int_0^T H_t d\tilde{S}_t$. Note that hedging in continuous time is an idealization. This is usually assumed as it becomes more convenient to deal with (stochastic) differential equations than with finite difference equations.

Black–Scholes replication

Here we assume some familiarity with stochastic analysis. However, this section is not needed for the rest of the book and therefore can be skipped (see however the expression of the Black–Scholes formula). We consider that S_t is modeled by a log-normal process under $\mathbb{P}^{\mathrm{hist}}$:

$$\frac{dS_t}{S_t} = \mu dt + \sigma dW_t^{\mathbb{P}^{\mathrm{hist}}}$$

$\mathcal{M}_\infty$ corresponds to the set of $\mathbb{Q}$-martingale measure equivalent to $\mathbb{P}^{\mathrm{hist}}$. From the Girsanov theorem (see e.g. [130]) , $\mathcal{M}_\infty$ reduces to a singleton $\{\mathbb{Q}^{\mathrm{BS}}\}$ under which

$$\frac{dS_t}{S_t} = rdt + \sigma dW_t^{\mathbb{Q}^{\mathrm{BS}}}$$

where $W^{\mathbb{Q}^{\mathrm{BS}}}$ is a $\mathbb{Q}^{\mathrm{BS}}$-Brownian motion with $\frac{d\mathbb{Q}^{\mathrm{BS}}}{d\mathbb{P}^{\mathrm{hist}}} = e^{-\frac{(\mu-r)^2 t}{2\sigma^2} - \frac{(\mu-r)}{\sigma}W_t^{\mathbb{P}^{\mathrm{hist}}}}$. We conclude that there is a unique arbitrage-free price (independent of μ - compare with formula (1.6)):

$$C = \mathbb{E}^{\mathbb{Q}^{\mathrm{BS}}}[e^{-rT}F_T]$$

We deduce also that the payoff can be dynamically hedged:

$$-e^{-rT}F_T + C + \int_0^T \partial_{S_t}\mathbb{E}^{\mathbb{Q}^{\mathrm{BS}}}[e^{-r(T-t)}F_T|S_t]d\tilde{S}_t = 0, \quad \mathbb{P}^{\mathrm{hist}} - \mathrm{a.s.}$$

Note that for a call payoff $F_T = (S_T - K)^+$, we obtain the Black–Scholes formula:

DEFINITION 1.11 Black–Scholes formula

$$C_{\mathrm{BS}}(\sigma^2 T, K) = S_0 N(d_+) - Ke^{-rT}N(d_-) \qquad (1.25)$$

with $d_{\pm} = -\dfrac{\ln \frac{Ke^{-rT}}{S_0}}{\sigma\sqrt{T}} \pm \dfrac{\sigma\sqrt{T}}{2}$. *Here* $N(x) = \int_{-\infty}^{x} e^{-\frac{y^2}{2}} \dfrac{dy}{\sqrt{2\pi}}$.

Chapter 2

Martingale optimal transport

Abstract   This chapter is at the core of this book. After reviewing key notions in OT, we introduce a constrained version where a martingale condition is imposed. This MOT is linked to robust hedging and pricing of options consistent with Vanillas. No knowledge of OT is required to read this chapter.

2.1 Optimal transport in a nutshell

As our approach in MOT follows closely the analysis in classical OT, we review in this section key notions in OT such as Monge–Kantorovich duality and Brenier's theorem. We give its interpretation in mathematical finance. As a consequence, instead of presenting the main results in the setup of Polish spaces X, we focus on the simple case $X = \mathbb{R}_+$, eventually $\mathbb{R}_+^d$, relevant in finance. Our main reference is Chapter 1, 2, 3 and 5 in [139]. The reader is asked to report to these chapters for detailed proofs (with weak assumptions).

2.1.1 Trading T-Vanilla options

We assume that T-Vanilla options on each asset are traded on the market. They are specified by a payoff $\lambda(S_T)$ at a maturity T. In practice, these Vanilla payoffs can be replicated by holding a strip of put/call T-Vanillas through the Taylor expansion formula [38]:

$$\lambda(S_T) = \lambda(S_0) + \lambda'(S_0)(S_T - S_0) + \int_0^{S_0} \lambda''(K)(K - S_T)^+ dK$$
$$+ \int_{S_0}^{\infty} \lambda''(K)(S_T - K)^+ dK$$

where $(K - S_T)^+$ (resp. $(S_T - K)^+$) is the payoff of a put (resp. call). Derivatives $\lambda''(K)$ are understood in the distribution sense. We then assume that the pricing operator $\Pi[\cdot]$ (used by market operators to value Vanillas) is linear meaning that

$$\Pi\left[\sum_i \lambda_i (S_T - K_i)^+\right] = \sum_i \lambda_i \Pi[(S_T - K_i)^+]$$

Moreover, from the no-arbitrage condition, we should have that

$$\Pi[1] = e^{-rT}, \quad \Pi[S_T] = S_0 \tag{2.1}$$

Also, still from the no-arbitrage condition, $\Pi[(S_T - K)^+]$ should be non-increasing, convex with respect to K and $\Pi[(S_T - K)^+] \geq (S_0 - Ke^{-rT})^+$.
From Riesz's representation theorem (with the additional requirement that the market price of a call option with strike K goes to 0 as $K \to \infty$), this implies that there exists a probability $\mathbb{P}^{\mathrm{mkt}}$ such that

$$C(K) \equiv \Pi[(S_T - K)^+] = \mathbb{E}^{\mathbb{P}^{\mathrm{mkt}}}[e^{-rT}(S_T - K)^+]$$

with $\mathbb{E}^{\mathbb{P}^{\mathrm{mkt}}}[e^{-rT}S_T] = S_0$.
Below and in the rest of the book, for the sake of simplicity, we take $r = 0$. This can be easily relaxed by including in the formulas below a multiplicative factor e^{-rT}.
From the linear property, the market price of the payoff $\lambda(S_T)$, inferred from market prices of put/call options, is

$$\Pi[\lambda(S_T)] = \mathbb{E}^{\mathbb{P}^{\mathrm{mkt}}}[\lambda(S_T)] = \lambda(S_0) + \int_0^{S_0} \lambda''(K)\mathbb{E}^{\mathbb{P}^{\mathrm{mkt}}}[(K - S_T)^+]dK$$

$$+ \int_{S_0}^{\infty} \lambda''(K)\mathbb{E}^{\mathbb{P}^{\mathrm{mkt}}}[(S_T - K)^+]dK \tag{2.2}$$

From the market value $C^i(K)$ of a call option with strike K and maturity T on asset i, one can deduce the T-marginal distributions by

$$\mathbb{P}^{\mathrm{mkt}}(S_T^i = K) = \partial_K^2 C^i(K), \quad i = 1, 2$$

This was first observed by [28]. The second-order derivative is defined almost everywhere as $C^i(K)$ is a convex function. In practice, market prices of put/call options (of strike K and maturity T) are block-coded through the implied volatility $\sigma_{\mathrm{BS}}(T, K)$:

DEFINITION 2.1 Implied volatility   *The implied volatility $\sigma_{\mathrm{BS}}(T, K)$ is defined as the number which, when put in the Black–Scholes formula (see Equation 1.25) for a call option with strike K and maturity T, reproduces the market price $C^{\mathrm{mkt}}(T, K)$: $\sigma_{\mathrm{BS}}(T, K)$ such that*

$$C^{\mathrm{mkt}}(T, K) = C_{\mathrm{BS}}(\sigma_{\mathrm{BS}}(T, K)^2 T, K)$$

As C_{BS} is monotone in σ, there is a unique solution.

2.1.2   Super-replication and Monge–Kantorovich duality

Let us start with two assets S_1 and S_2 evaluated at the same maturity T and consider a payoff $c(s_1, s_2)$ depending on these two underlyings. Here in comparison with the last section, we have skipped the index T for the asset price for the ease of notation: $S_1 \equiv S_T^1$ and $S_2 \equiv S_T^2$. We have the following assumption on c:

Assumption 3 : *Let $c : \mathbb{R}_+^2 \to [-\infty, \infty)$ be a continuous function such that*

$$c^+(s_1, s_2) \leq K \cdot (1 + s_1 + s_2) \tag{2.3}$$

on $(\mathbb{R}_+)^2$ for some constant K.

Below we denote $\mathbb{P}^1$ (resp. $\mathbb{P}^2$) the distribution implied from T-Vanillas on S_1 (resp. S_2), see the previous section. From (2.1), the first moment of $\mathbb{P}^1$ (resp. $\mathbb{P}^2$) is finite equal to the spot value at $t = 0$: S_1^0 (resp. S_2^0).
The model-independent super-replication price (consistent with T-Vanillas on assets 1 and 2) is then defined as

DEFINITION 2.2

$$\mathrm{MK}_2 \equiv \inf_{\mathcal{P}^*(\mathbb{P}^1, \mathbb{P}^2)} \mathbb{E}^{\mathbb{P}^1}[\lambda_1(S_1)] + \mathbb{E}^{\mathbb{P}^2}[\lambda_2(S_2)] \tag{2.4}$$

where $\mathcal{P}^*(\mathbb{P}^1, \mathbb{P}^2)$ is the set of all functions $(\lambda_1, \lambda_2) \in \mathrm{L}^1(\mathbb{P}^1) \times \mathrm{L}^1(\mathbb{P}^2)$ such that

$$\lambda_1(s_1) + \lambda_2(s_2) \geq c(s_1, s_2) \tag{2.5}$$

for $\mathbb{P}^1$-almost all $s_1 \in \mathbb{R}_+$ and $\mathbb{P}^2$-almost all $s_2 \in \mathbb{R}_+$.

One can show that the infimum in (2.4) is attained and the value of the infimum does not change if one restricts the definition of $\mathcal{P}^*(\mathbb{P}^1, \mathbb{P}^2)$ to those functions which are bounded and continuous (see Theorem 1.3 in [139]). As a consequence, the inequality $\lambda_1(s_1) + \lambda_2(s_2) \geq c(s_1, s_2)$ holds true for all $(s_1, s_2) \in \mathbb{R}_+^2$. It is clear that this infimum is not unique as if $(\lambda_1, \lambda_2) \in \mathcal{P}^*(\mathbb{P}^1, \mathbb{P}^2)$, then $(\lambda_1 + C, \lambda_2 - C) \in \mathcal{P}^*(\mathbb{P}^1, \mathbb{P}^2)$ corresponding to shifting the European payoffs by a constant.
This static super-replication strategy consists in holding Vanilla payoffs $\lambda_1(s_1)$ and $\lambda_2(s_2)$ with $(t = 0)$ market prices $\mathbb{E}^{\mathbb{P}^1}[\lambda_1(S_1)]$, $\mathbb{E}^{\mathbb{P}^2}[\lambda_2(S_2)]$ such that the value of this portfolio $\lambda_1(s_1) + \lambda_2(s_2)$ at maturity is greater than or equal to the payoff $c(s_1, s_2)$.

Optimal transport formulation

The linear program (2.4), called the Monge–Kantorovich formulation of OT, can be dualized by introducing a Kuhn–Tucker multiplier (i.e., a positive measure on $\mathbb{R}_+^2$) associated to the inequality (2.5):

THEOREM 2.1

$$\text{MK}_2 = \sup_{\mathbb{P} \in \mathcal{P}(\mathbb{P}^1, \mathbb{P}^2)} \mathbb{E}^{\mathbb{P}}[c(S_1, S_2)] \tag{2.6}$$

with $\mathcal{P}(\mathbb{P}^1, \mathbb{P}^2) = \{\mathbb{P} : S_1 \overset{\mathbb{P}}{\sim} \mathbb{P}^1, S_2 \overset{\mathbb{P}}{\sim} \mathbb{P}^2\}.$

This dual expression coincides with the Kantorovich formulation of OT. This result is known as the Kantorovich duality. Note that the OT is usually presented in textbooks with an inf instead of a sup. The (dual) optimization MK_2 consists in maximizing the cost function $\mathbb{E}^{\mathbb{P}}[c(S_1, S_2)]$ over the (convex) set of joint measures $\mathbb{P}$ with marginals $\mathbb{P}^1$ and $\mathbb{P}^2$. Notice that $\mathcal{P}(\mathbb{P}^1, \mathbb{P}^2)$ is non-empty as it contains the trivial coupling measure $\mathbb{P}^1 \otimes \mathbb{P}^2$, the product measure under which S_1 and S_2 are independent.

PROOF (Sketch), see [139] Theorem 1.3 We quickly derive the dual formulation by using a minimax principle, which copycats the proof of Theorem 1.1 (the reader should refer to this proof for additional details). It consists in writing the constrained primal problem (2.4) into an unconstrained problem by introducing a Kuhn–Tucker multiplier $\mathbb{P}$:

$$\text{MK}_2 = \inf_{(\lambda_i(\cdot))_{1 \leq i \leq 2}} \sup_{\mathbb{P} \in \mathcal{M}_+} \mathbb{E}^{\mathbb{P}}[c(S_1, S_2) - \sum_{i=1}^{2} \lambda_i(S_i)] + \sum_{i=1}^{2} \mathbb{E}^{\mathbb{P}^i}[\lambda_i(S_i)]$$

where $\mathcal{M}_+$ is the set of positive measure on $\mathbb{R}_+^2$. The sup over $\mathbb{P}$ will be ∞ except if the constraint (2.5) is satisfied for all $(s_1, s_2) \in \mathbb{R}_+^2$ (in this case the sup vanishes). The minimax principle consists then in switching the inf and sup operations:

$$\text{MK}_2 = \sup_{\mathbb{P} \in \mathcal{M}_+} \mathbb{E}^{\mathbb{P}}[c(S_1, S_2)] + \inf_{(\lambda_i(\cdot))_{1 \leq i \leq 2}} \sum_{i=1}^{2} \left(\mathbb{E}^{\mathbb{P}^i}[\lambda_i(S_i)] - \mathbb{E}^{\mathbb{P}}[\lambda_i(S_i)] \right)$$

The infimum over λ_i will be $-\infty$ except if $\mathbb{P}$ has the marginals $\mathbb{P}^i$. This reproduces our expression (2.6) for MK_2.   ꠸

For a continuous cost function, as the set $\mathcal{P}(\mathbb{P}^1, \mathbb{P}^2)$ is weakly compact (i.e., with respect to the convergence in distribution) from Prokhorov's theorem[1], the supremum over $\mathcal{P}(\mathbb{P}^1, \mathbb{P}^2)$ is attained. This means that there exists $\mathbb{P}^* \in \mathcal{P}(\mathbb{P}^1, \mathbb{P}^2)$ (not necessarily unique) such that $\text{MK}_2 = \mathbb{E}^{\mathbb{P}^*}[c(S_1, S_2)]$.

---

[1] Let K_ϵ^1 and K_ϵ^2 be two compacts such that $\mathbb{P}^1(S_1 \in \mathbb{R}_+ \setminus K_1^\epsilon) \leq \epsilon$ and $\mathbb{P}^1(S_2 \in \mathbb{R}_+ \setminus K_2^\epsilon) \leq \epsilon$. Use that $\mathbb{P}((S_1, S_2) \in \mathbb{R}_+^2 \setminus K_\epsilon^1 \times K_\epsilon^2) \leq \mathbb{P}^1(S_1 \in \mathbb{R}_+ \setminus K_1^\epsilon) + \mathbb{P}^2(S_2 \in \mathbb{R}_+ \setminus K_2^\epsilon) \leq 2\epsilon$ for proving the tightness and therefore the relatively compactness from Prokhorov's theorem (see Remark 1.4). We conclude as $\mathcal{P}(\mathbb{P}^1, \mathbb{P}^2)$ is a closed set for the weak topology.

REMARK 2.1 Weak/strong duality Theorem 2.1 corresponds to a strong duality result. From the inequality (1.10), one can prove the weak duality result

$$\text{MK}_2 \geq \sup_{\mathbb{P} \in \mathcal{P}(\mathbb{P}^1, \mathbb{P}^2)} \mathbb{E}^{\mathbb{P}}[c(S_1, S_2)] \qquad (2.7)$$

Here we don't need to invoke a minimax argument. Alternatively, this can be obtained by taking the expectation of $\lambda_1(S_1) + \lambda_2(S_2) \geq c(S_1, S_2)$ for all $\mathbb{P} \in \mathcal{P}(\mathbb{P}^1, \mathbb{P}^2)$:

$$\mathbb{E}^{\mathbb{P}^1}[\lambda_1(S_1)] + \mathbb{E}^{\mathbb{P}^2}[\lambda_2(S_2)] \geq \mathbb{E}^{\mathbb{P}}[c(S_1, S_2)]$$

◻

For further reference, we quote the following proposition that proves that the infimum is attained by a pair (λ, λ^*) of bounded continuous (i.e., C_b) c-concave functions, defined below:

PROPOSITION 2.1

$$\text{MK}_2 = \inf_{\lambda \in C_b} \mathbb{E}^{\mathbb{P}^1}[\lambda^*(S_1)] + \mathbb{E}^{\mathbb{P}^2}[\lambda(S_2)] \qquad (2.8)$$

with $\lambda^*(s_1) \equiv \sup_{s_2 \in \mathbb{R}_+} \{c(s_1, s_2) - \lambda(s_2)\}$ *the c-concave transform of λ.*

PROOF (Sketch), see [139] Theorem 2.9 For all $(\lambda_1, \lambda_2) \in \mathcal{P}^*(\mathbb{P}^1, \mathbb{P}^2)$, we have $\lambda_1 \geq c - \lambda_2$ for all s_2 and therefore $\lambda_1(s_1) \geq \lambda_2^*(s_1)$. As the cost $\mathbb{E}^{\mathbb{P}^1}[\lambda_1] + \mathbb{E}^{\mathbb{P}^2}[\lambda_2]$ is linear with respect to λ_1, the infimum over λ_1 is reached for $\lambda_1 = \lambda_1^*$. We can then conclude (we need to ensure that $\lambda_2^* \in L^1(\mathbb{P}^1)$). ◻

Note that all references in OT present first the dual (2.6) and then the primal formulation (2.4). In the context of mathematical finance, the primal (2.4), being interpreted as the robust super-replication price of c, has a clear financial interpretation and therefore is presented first.

Below, we give an example of applications of OT in the construction of arbitrage-free currency cross-rate implied volatility, drawn from [93].

Example 2.1 Illiquid currency cross-rate implied volatility
We consider the arbitrage-freeness of a currency cross-rate market implied volatility (see Definition 2.1) using the market implied volatilities of two related exchange rates. More precisely, let us consider Vanilla payoffs on the exchange rates linking the following three currencies: JPY, US dollars and Australian dollars. We denote the dollar price of one Japanese Yen at maturity T by $S_1 \equiv S_T^{\$/\text{JPY}}$ and likewise the dollar price of one Australian dollar

by $S_2 \equiv S_T^{\$/\text{AUD}}$. The value of one JPY in Australian dollars is then given by $S_3 \equiv S_T^{\text{JPY}/\text{AUD}} = S_2/S_1$. Hence, a cross-rate call option is equivalent to an option to exchange one asset for the other asset and its payoff equals $(S_2 - KS_1)^+$. The arbitrage-free market price of this payoff can be written as (see Section 2.1.1)

$$\mathbb{E}^{\mathbb{P}^{\text{mkt}}}[S_1 \left(\frac{S_2}{S_1} - K \right)^+]$$

for a two-dimensional probability $\mathbb{P}^{\text{mkt}}(ds_1, ds_2)$. By doing a change of measure, this can be written as

$$S_1^0 \mathbb{E}^{\mathbb{P}^3}[(S_3 - K)^+]$$

where $\mathbb{P}^3$, defined by $\frac{d\mathbb{P}^3}{d\mathbb{P}^{\text{mkt}}} = \frac{S_1}{S_1^0}$, can now be interpreted as the distribution of $S_3 = S_2/S_1$. By linearity, the market price of the payoff $S_1 \lambda_3(S_3)$ is $S_1^0 \mathbb{E}^{\mathbb{P}^3}[\lambda_3(S_3)]$.

The triangle consisting in T-Vanillas on S_1, S_2 and S_3 is arbitrage-free if the bound MK_3 defined below is zero:

$$\text{MK}_3 = \inf_{\mathcal{P}^*(\mathbb{P}^1,\mathbb{P}^2,\mathbb{P}^3)} \mathbb{E}^{\mathbb{P}^1}[\lambda_1(S_1)] + \mathbb{E}^{\mathbb{P}^2}[\lambda_2(S_2)] + S_1^0 \mathbb{E}^{\mathbb{P}^3}[\lambda_3(S_3)]$$

where $\mathcal{P}^*(\mathbb{P}^1, \mathbb{P}^2, \mathbb{P}^3)$ is the set of continuous functions in $L^1(\mathbb{P}^1) \times L^1(\mathbb{P}^2) \times L^1(\mathbb{P}^3)$ such that

$$\lambda_1(s_1) + \lambda_2(s_2) + s_1 \lambda_3(s_2/s_1) \geq 0$$

for all $(s_1, s_2) \in \mathbb{R}_+^2$. Indeed $\text{MK}_3 \leq 0$ as $(\lambda_1 = 0, \lambda_2 = 0, \lambda_3 = 0) \in \mathcal{P}^*(\mathbb{P}^1, \mathbb{P}^2, \mathbb{P}^3)$ and if $\text{MK}_3 < 0$, then the arbitrage can be locked in by buying the static portfolio $\lambda_1(s_1) + \lambda_2(s_2) + s_1 \lambda_3(s_2/s_1)$, generating a nonnegative profit, at the negative price MK_3. This would indicate an arbitrage opportunity in the market of T-Vanillas on S_1, S_2 and S_3. MK_3 is equivalent to (see Equation 2.8)

$$\text{MK}_3 = \inf_{\lambda_1 \in L^1(\mathbb{P}^1), \lambda_2 \in L^1(\mathbb{P}^2)} \mathbb{E}^{\mathbb{P}^1}[\lambda_1(S_1)] + \mathbb{E}^{\mathbb{P}^2}[\lambda_2(S_2)] - S_1^0 \mathbb{E}^{\mathbb{P}^3}[\lambda_3^*(S_3)]$$

Here λ_3^* is the (multiplicative) inf-convolution of λ_1 and λ_2:

$$\lambda_3^*(s_3) \equiv \inf_{s_1 \in \mathbb{R}_+} \frac{1}{s_1}\{\lambda_1(s_1) + \lambda_2(s_3 s_1)\}$$

By assuming that a finite number of T-calls can be traded on the market, MK_3 defines a finite-dimensional linear program that can be solved with a simplex algorithm. A triangle arbitrage is detected if $\text{MK}_3 \neq 0$, meaning by duality that the subset of probabilities $\mathbb{P}(dx, dy)$ on $\mathbb{R}_+^2$, $\mathcal{P}(\mathbb{P}^1, \mathbb{P}^2, \mathbb{P}^3) \equiv \{\mathbb{P} : S_1 \overset{\mathbb{P}}{\sim} \mathbb{P}^1, S_2 \overset{\mathbb{P}}{\sim} \mathbb{P}^2, \int x^2 \mathbb{P}(x, Kx)dx = S_1^0 \mathbb{P}^3(K) \; \forall \; K \in \mathbb{R}_+\}$, is empty. We have used that $\mathbb{E}^{\mathbb{P}}[S_1(S_2/S_1 - K)^+] = S_1^0 \mathbb{E}^{\mathbb{P}^3}[(S_3 - K)^+]$ and therefore $\mathbb{E}^{\mathbb{P}}[S_1^2 \delta(S_2 - KS_1)] = S_1^0 \mathbb{E}^{\mathbb{P}^3}[\delta(S_3 - K)^+]$. $\quad\square$

2.1.3 Formulation in $\mathbb{R}_+^d$ and multi-dimensional marginals

The MK formulation and its dual expression remain valid when S_1 and S_2 are two random variables in $\mathbb{R}_+^d$. The interpretation in mathematical finance goes as follows: let us consider a payoff $c(s_1, s_2)$ depending on two groups (s_1, s_2), each composed of d assets. The first group is $(s_1^1, \ldots, s_1^d) \in \mathbb{R}_+^d$. Knowing the distribution of $S_1 \in \mathbb{R}_+^d$ is equivalent to knowing (at $t = 0$) the market values of all basket options $\mathbb{E}^{\mathbb{P}^1}[(S_1 \cdot \omega - K)^+]$ for all $K \in \mathbb{R}$ and for all $\omega \in \mathbb{R}^d$. This equivalence can be seen by observing that basket option prices fix the Laplace transform of S_1: $\mathbb{E}^{\mathbb{P}^1}[e^{\omega \cdot S_1}]$. Although basket options are liquid only for some particular values of the weight ω (and K), the values $\mathbb{E}^{\mathbb{P}^1}[(S_1 \cdot \omega - K)^+]$ can be however fixed by assuming a correlation structure (more precisely a copula, denoted co below) between the variables $(S_1^1, \ldots, S_1^d)$. For example, the first group of assets (resp. second) belongs to the same financial sector and can therefore be assumed to be strongly correlated. This is not the case for the correlation structures between S_1 and S_2 which belong to two different groups and for which the correlation information is difficult to obtain. This is found through our OT formulation. By definition of the copula co, we impose that

$$\mathbb{E}^{\mathbb{P}^1}[\lambda_1(S_1)] \equiv \mathbb{E}[\lambda_1(F_1^{-1}(U_1), \ldots F_d^{-1}(U_d))co(U_1, \ldots, U_d)]$$

where $(U_i)_{1 \le i \le d}$ are d independent uniform random variables and F_i is the cumulative distribution of S_1^i implied from T-Vanilla options on S_1^i. Note that our discussion can be extended when $S_1 \in \mathbb{R}_+^d$ and $S_2 \in \mathbb{R}_+^{d^*}$ with $d \ne d^*$.

2.1.4 Fréchet–Hoeffding solution

Under the so-called Spence–Mirrlees condition, $c_{12} \equiv \partial_{s_1 s_2} c > 0$, OT (2.6) can be solved explicitly. Let F_1, F_2 denote the cumulative distribution functions of $\mathbb{P}^1$ and $\mathbb{P}^2$. For the sake of simplicity, we will assume that $\mathbb{P}^1$ does not give mass to points and $c \in C^2$.

THEOREM 2.2
Under $c_{12} > 0$,
(i): The optimal measure $\mathbb{P}^*$ has the form

$$\mathbb{P}^*(ds_1, ds_2) = \delta_{T(s_1)}(ds_2)\,\mathbb{P}^1(ds_1)$$

with T the forward image of the measure $\mathbb{P}^1$ onto $\mathbb{P}^2$: $T(x) = F_2^{-1} \circ F_1(x)$.
(ii): The optimal upper bound is given by

$$\mathrm{MK}_2 = \int_0^1 c(F_1^{-1}(u), F_2^{-1}(u))du$$

This optimal bound can be attained by a static hedging strategy consisting in holding European payoffs $\lambda_1 \in \mathrm{L}^1(\mathbb{P}^1)$, $\lambda_2 \in \mathrm{L}^1(\mathbb{P}^2)$ with market prices

$\mathbb{E}^{\mathbb{P}^1}[\lambda_1(S_1)]$ *and* $\mathbb{E}^{\mathbb{P}^2}[\lambda_2(S_2)]$:

$$\mathrm{MK}_2 = \mathbb{E}^{\mathbb{P}^1}[\lambda_1(S_1)] + \mathbb{E}^{\mathbb{P}^2}[\lambda_2(S_2)]$$

with

$$\lambda_2(x) = \int_0^x c_2(T^{-1}(y), y) dy, \quad \lambda_1(x) = c(x, T(x)) - \lambda_2(T(x))$$

The value of this static European portfolio super-replicates the payoff at maturity:

$$\lambda_1(s_1) + \lambda_2(s_2) \geq c(s_1, s_2), \quad \forall \, (s_1, s_2) \in \mathbb{R}_+^2$$

T is refereed as the Brenier map (or Fréchet–Hoeffding).
Note that the above theorem requires additional conditions on c in order to guarantee the integrability conditions $\lambda_1 \in L^1(\mathbb{P}^1)$ and $\lambda_2 \in L^1(\mathbb{P}^2)$.

PROOF   The proof relies heavily on our duality result (more precisely, the weak duality (2.7) is enough).
(i): Our guess for the optimal probability $\mathbb{P}^*$ is

$$\mathbb{P}^*(ds_1, ds_2) = \delta_{T(s_1)}(ds_2) \, \mathbb{P}^1(ds_1) \; : \; T(s_1) = F_2^{-1} \circ F_1(s_1)$$

As $\mathbb{P}^* \in \mathcal{P}(\mathbb{P}^1, \mathbb{P}^2)$, this leads from the weak duality to a lower bound $\underline{D} \leq \mathrm{MK}_2$ with

$$\underline{D} = \int_0^1 c(F_1^{-1}(u), F_2^{-1}(u)) du$$

(ii): We will derive our guess for λ_1 and λ_2. If the primal is attained (this is always the case, see [139] Theorem 1.3 but one does not need to use this result) - say for λ_1 and λ_2 - we should have if $\mathbb{P}^*$ is our optimal probability

$$\mathbb{E}^{\mathbb{P}^*}[c - \lambda_1(S_1) - \lambda_2(S_2)] = \mathbb{E}^{\mathbb{P}^*}[c] - \mathbb{E}^{\mathbb{P}^1}[\lambda_1(S_1)] - \mathbb{E}^{\mathbb{P}^2}[\lambda_2(S_2)] = 0$$

which implies as $\lambda_1(s_1) + \lambda_2(s_2) - c(s_1, s_2) \geq 0$:

$$\lambda_1(s_1) + \lambda_2(s_2) = c(s_1, s_2), \quad \mathbb{P}^* - \text{a.s.}$$

This means

$$\lambda_1(x) + \lambda_2(T(x)) = c(x, T(x)), \quad \forall \, x \in \mathbb{R}_+ \tag{2.9}$$

Additionally, λ_1 should be the c-concave transform of λ_2 (see Proposition 2.1):

$$\lambda_1(s_1) = \sup_{s_2}\{c(s_1, s_2) - \lambda_2(s_2)\}$$

The supremum should be reached at the point $s_2 = T(s_1)$ meaning that

$$c_2(s_1, T(s_1)) - \lambda_2'(T(s_1)) = 0 \tag{2.10}$$

and $c_{22}(s_1, T(s_1)) - \lambda_2''(T(s_1)) \leq 0$.

Equations (2.9, 2.10) give our guess for the primal problem:

$$\lambda_2(x) = \int_0^x c_2(T^{-1}(y), y) dy$$

$$\lambda_1(x) = c(x, T(x)) - \lambda_2(T(x))$$

(iii): $(\lambda_1, \lambda_2) \in \mathcal{P}^*(\mathbb{P}^1, \mathbb{P}^2)$ if and only if $c_{12} > 0$ as

$$c_2(s_1, x) - \lambda_2'(x) = \int_{T^{-1}(x)}^{s_1} c_{12}(y, x) dy$$

(and therefore $c(s_1, s_2) - \lambda_2(s_2)$ attains its maximum only at $s_2 = T(s_1)$). This gives the upper bound $MK_2 \leq \overline{P}$ with

$$\overline{P} = \mathbb{E}^{\mathbb{P}^1}[\lambda_1(S_1)] + \mathbb{E}^{\mathbb{P}^2}[\lambda_2(S_2)]$$

From the (weak) duality, we get

$$\underline{D} \leq MK_2 \leq \overline{P}$$

We conclude the proof as a straightforward computation gives that $\overline{P} = \underline{D}$. ☐

REMARK 2.2 This result indicates that the upper bound is obtained by perfectly correlating the two assets S_1 and S_2. For example, if $F_1(S_1) = N(G_1)$ and $F_2(S_2) = N(G_2)$ where G_1 and G_2 are two correlated Gaussian variables with correlation ρ, $\mathbb{E}^{\mathbb{P}}[c(S_1, S_2)]$ is an increasing function in ρ under the condition $c_{12} > 0$. ☐

Example 2.2 Upper bound, $c(s_1, s_2) = (s_1 - K_1)^+ 1_{s_2 > K_2}$

By applying Fréchet–Hoeffding solution, the upper bound is attained by

$$MK_2 = \int_{\max(F_1(K_1), F_2(K_2))}^1 (F_1^{-1}(u) - K_1) du$$

with

$$\lambda_2(x) = \left(F_1^{-1} \circ F_2(K_2) - K_1\right)^+ 1_{x > K_2}$$

$$\lambda_1(x) = (x - K_1)^+ 1_{F_2^{-1} \circ F_1(x) > K_2} - \left(F_1^{-1} \circ F_2(K_2) - K_1\right)^+ 1_{F_2^{-1} \circ F_1(x) > K_2}$$

In the particular case where $F_2 = F_1$, we have

$$\lambda_2(x) = (K_2 - K_1)^+ 1_{x > K_2}$$

$$\lambda_1(x) = (x - K_1)^+ 1_{x > K_2} - (K_2 - K_1)^+ 1_{x > K_2}$$

☐

Monge's solution

The Fréchet–Hoeffding solution (or more generally Brenier's solution in $\mathbb{R}^d$- see next section) is also the optimal solution of the Monge OT defined by

$$\mathrm{M}_2 \equiv \sup_{\mathcal{P}^{\mathrm{Monge}}(\mathbb{P}^1, \mathbb{P}^2)} \mathbb{E}^{\mathbb{P}}[c(S_1, S_2)]$$

with the set

$$\mathcal{P}^{\mathrm{Monge}}(\mathbb{P}^1, \mathbb{P}^2) \equiv \{\mathbb{P}_T \in \mathcal{P}(\mathbb{P}^1, \mathbb{P}^2) \ : \ \mathbb{P}_T(ds_1, ds_2) = \delta_{T(s_1)}(ds_2)\, \mathbb{P}^1(s_1)\}$$

Note that this problem is more involved in $\mathbb{R}^d$ as the constraint on T is nonlinear (see section on Monge–Ampère equation). It is therefore simpler to start with the (relaxed) Monge–Kantorovich formulation MK_2 and then show that it coincides with the Monge problem: $\mathrm{MK}_2 = \mathrm{M}_2$.

From a financial point of view, only the primal formulation of MK_2 has a clear interpretation.

2.1.5   Brenier's solution

The Fréchet–Hoeffding solution has been generalized in $\mathbb{R}^d$ by Brenier [29] first in the case of a quadratic cost function and then extended to concave payoff $c = c(s_1 - s_2)$ by Gangbo and McCann [79] and others:

THEOREM 2.3 Brenier - $c(s_1, s_2) = -|s_1 - s_2|^2/2$

(i): *If $\mathbb{P}^1$ has no atoms, then there is a unique optimal $\mathbb{P}^*$, which is a Monge solution:*

$$\mathbb{P}^* = \delta_{T(s_1)}(ds_2)\, \mathbb{P}^1(s_1) \tag{2.11}$$

with $T = \nabla\lambda_1$. $\nabla\lambda_1$ is the unique gradient of a convex function λ_1.

(ii): *The optimal bound is attained by a static hedging strategy with $\lambda_2(x) = c(x, T(x)) - \lambda_1(x)$ and λ_1 uniquely specified by*

$$(\nabla\lambda_1)\#\mathbb{P}^1 = \mathbb{P}^2$$

The notation $T\#\mathbb{P}^1 = \mathbb{P}^2$ means that for all $U \in \mathrm{L}^1(\mathbb{P}^2)$:

$$\mathbb{E}^{\mathbb{P}^1}[U(T(S_1))] = \mathbb{E}^{\mathbb{P}^2}[U(S_2)] \tag{2.12}$$

If T is differentiable, this condition reads

$$|\det \nabla T|\mathbb{P}^2(T(x)) = \mathbb{P}^1(x)$$

This theorem has been generalized to a strictly concave, superlinear[2] cost function $c(s_1, s_2) = c(s_1 - s_2)$. The Brenier map is then

$$T(x) = x - \nabla c^*(\nabla \lambda_1(x)) \tag{2.13}$$

for some c-concave function λ_1 which is uniquely fixed by the requirement $T_\# \mathbb{P}^1 = \mathbb{P}^2$. Here $c^* \equiv \inf_x \{p.x - c(x)\}$ is the Legendre transform of c.

PROOF (Sketch), cost function $c(s_1 - s_2)$ strictly concave
Our proof follows closely our proof in $d = 1$ (see Theorem 2.2).
(i): On the dual side, by assuming that we have a feasible solution of the form (2.11), we obtain a lower bound $\text{MK}_2 \geq \mathbb{E}^{\mathbb{P}^1}[c(S_1, T(S_1))]$.
(ii): On the primal side, the optimal λ_1 and λ_2 should satisfy

$$\lambda_1(x) + \lambda_2(T(x)) = c(x - T(x)) \tag{2.14}$$
$$\nabla \lambda_1(x) = (\nabla c)(x - T(x)) \tag{2.15}$$

Note that the Legendre transform of c is $c^*(p) = p.x(p) - c(x(p))$ with $x(p)$ the unique solution of $p = \nabla c(x(p))$. By differentiating $c^*(p)$ with respect to p, we obtain $\nabla c^*(p) = x(p)$. This implies that Equation (2.15) can be solved as

$$x - T(x) = \nabla c^*(\nabla \lambda_1(x))$$

This fixes uniquely the scalar function λ_1 (and therefore T) with the requirement $\mathbb{P}^2(T(x)) \det(\nabla T) = \mathbb{P}^1(x)$ (see Monge–Ampère equation below). Under the concavity condition on c, one can show that λ_1 and λ_2 are feasible solutions of the primal problem and we obtain an upper bound : $\text{MK}_2 \leq \mathbb{E}^{\mathbb{P}^1}[\lambda_1(S_1)] + \mathbb{E}^{\mathbb{P}^2}[\lambda_2(S_2)]$. We conclude from the (weak) duality as

$$\mathbb{E}^{\mathbb{P}^1}[c(S_1, T(S_1))] \leq \text{MK}_2 \leq \mathbb{E}^{\mathbb{P}^1}[\lambda_1(S_1)] + \mathbb{E}^{\mathbb{P}^2}[\lambda_2(S_2)]$$

and $\mathbb{E}^{\mathbb{P}^1}[\lambda_1(S_1)] + \mathbb{E}^{\mathbb{P}^2}[\lambda_2(S_2)] = \mathbb{E}^{\mathbb{P}^1}[c(S_1, T(S_1))]$. ▯

Example 2.3 Wrong-way risk in CVA
Counterparty risk is the risk that a counterparty may default and fail to make future payments. CVA is defined as

$$\text{CVA} \equiv \mathbb{E}[1_{\tau < T} E_\tau]$$

where τ modeled the default of the counterparty and E_τ our exposure at default. E_τ is equal to a small fraction $(1 - R)$ of the positive part of the market-to-market MtM_τ: $E_\tau = (1 - R)\text{MtM}_\tau^+$. The market model, used by

---

[2] $\lim_{|s| \to \infty} |\frac{c(s)}{s}| = \infty$.

the bank for valuating the positive exposure, generates paths of discounted positive exposure at some dates $(t_i)_{i=1,...,n}$. CVA is then approximated by

$$\text{CVA} \approx \sum_{i=1}^{n} \mathbb{E}[1_{\tau \in [t_{i-1}, t_i]} E_{t_i}]$$

This can be put in a vector-valued form $\mathbb{E}[\tau \cdot E]$ where $E = (E_{t_1}, \ldots E_{t_n})$ and $\tau = (1_{\tau \in [t_0, t_1]}, \ldots 1_{\tau \in [t_{n-1}, t_n]})$. Knowing the marginals of E (as produced by our market model) and the marginals of τ^3, we would like to find the (optimal) bounds on the cost $\tau.E$. These bounds measure the so-called wrong-way risk and are precisely given by Brenier's solution for the quadratic cost as $2\tau.E = -(\tau - E)^2 + \tau^2 + E^2$. The computation of these bounds (using the MK formulation) is explored in [84].   ▯

Derivation of the Monge–Ampère equation

If we assume that $\mathbb{P}^1$ and $\mathbb{P}^2$ are absolutely continuous with respect to the Lebesgue measure, $\mathbb{P}^1(dx) \equiv p_1(x)dx$, $\mathbb{P}^2(dx) \equiv p_2(x)dx$, and $T \in C^1$, this implies that $u \equiv \lambda_1$ is then solution of a Monge–Ampère equation. The derivation is quite straightforward and is reported here. By differentiating Equation (2.13), we get

$$\nabla T = \text{Id} - \nabla(\nabla c^*(\nabla u))$$

By taking the determinant on both sides, and by noting that the condition (2.12) reads $\det(\nabla T)p_2(T(x)) = p_1(x)$, we get the generalized Monge–Ampère equation (Monge–Ampère equation for the quadratic cost)

$$\det\left(\text{Id} - \nabla(\nabla c^*(\nabla u))\right) = \frac{p_1(x)}{p_2(T(x))}$$

Note that the Monge–Ampère partial differential equation (in short PDE) arises in mathematical finance - see for example the problem of quantile hedging [25] - through the following stochastic representation

$$\sup_{\sigma \in S_d^+, \det(\sigma)=1} \text{Tr}[\sigma \nabla^2 u] + f(x) = 0$$

$$= n \det(\nabla^2 u)^{\frac{1}{n}} + f(x)$$

where S_d^+ denotes the set of nonnegative symmetric d-dimensional matrices[4]. This corresponds to an (infinite-horizon) Hamilton–Jacobi–Bellman (HJB) control problem in a multi-asset unbounded uncertain volatility model.

---

[3]Inferred from market prices of credit-default swaps.

[4]We use the identity $\sup_{(a_i)_{i=1,...,n}>0, \prod_{i=1}^{n} a_i=1} \sum_{i=1}^{n} a_i \lambda_i = n(\prod_{i=1}^{n} \lambda_i)^{\frac{1}{n}}$.

2.1.6 Axiomatic construction of marginals: Stieltjes moment problem

We have explained previously that marginals $\mathbb{P}^i$ can be inferred from market values of T-Vanilla call/put options. However, in practice, only a finite number of strikes are quoted and therefore these liquid prices need to be interpolated and extrapolated in order to imply the marginals $\mathbb{P}^i$ (supported in $\mathbb{R}_+$). We report here how this can be achieved. This problem can be framed as Stieltjes moment problem. By construction, our T-marginal should belong to the infinite-dimensional convex set

$$\mathcal{M} = \{\mathbb{P} : \mathbb{E}^{\mathbb{P}}[S_T] = S_0, \quad \mathbb{E}^{\mathbb{P}}[(S_T - K_i)^+] = C(K_i) \equiv c_i, \quad i = 1, \ldots n\}$$

$\mathcal{M}$ is relatively compact from Prokhorov's theorem (See Remark 1.4). For instance, we add the technical assumption that the elements in $\mathcal{M}$ should also be compactly supported in the interval $[0, S_{\max}]$ with $S_{\max}$ large in order to get that $\mathcal{M}$ is compact. This implies that from Krein–Milman's theorem, this set can be reconstructed from its extremal points $\text{Ext}(\mathcal{M})$:

$$\mathcal{M} = \overline{\text{Conv}(\text{Ext}(\mathcal{M}))}$$

Furthermore, from Choquet's theorem, one can show that all arbitrage-free prices $C(K)$ can be obtained by linearly combining extremal points. They are supported by a probability measure μ on $\text{Ext}(\mathcal{M})$ (probability on probability space!) and for all K,

$$C(K) = \int_{\text{Ext}(\mathcal{M})} \mathbb{E}^{\mathbb{P}}[(S_T - K)^+] d\mu(\mathbb{P})$$

Enumerating all the extremal points (and therefore elements in $\mathcal{M}$) is a difficult task. We follow a different route. A canonical point of $\mathcal{M}$ can be obtained by minimising a convex lower semi-continuous functional $\mathcal{F}$:

$$I \equiv \inf_{\mathbb{P} \in \mathcal{M}} \mathcal{F}(\mathbb{P}) = \mathcal{F}(\mathbb{P}^*_{c_1, \ldots, c_n}), \quad \mathbb{P}^*_{c_1, \ldots, c_n} \in \mathcal{M} \qquad (2.16)$$

Choice for $\mathcal{F}$

What is the best choice for $\mathcal{F}$? Here we list some criteria (which still generates infinite number of solutions): (1) $\mathbb{P}^*$ should be continuous with respect to $c_1, \ldots c_n$. (2) If $c_i \equiv c_i^{\text{BS}}(\sigma)$ are drawn from a log-normal distribution $\mathbb{P}^{\text{BS}}(\sigma)$ with a volatility σ (i.e., Black–Scholes distribution), then $\mathbb{P}^*_{c_1^{\text{BS}}(\sigma), \ldots, c_n^{\text{BS}}(\sigma)} = \mathbb{P}^{\text{BS}}(\sigma), \quad \forall \sigma, \quad \forall n \geq 0$.
As an example, satisfying the above criteria, we take

$$\mathcal{F}_1(\mathbb{P}|\mathbb{P}^0) = \int p^0(x) f\left(\frac{p(x)}{p^0(x)}\right) dx, \quad f \text{ positive, convex}, \quad f(1) = 0$$

$$(2.17)$$

$$= \infty, \quad \mathbb{P} \text{ not absolutely continuous w.r.t. } \mathbb{P}^0$$

We choose also for $\mathbb{P}^0(dx) = p^0(x)dx$ a log-normal distribution $\mathbb{P}^{BS}(\sigma)$ with a volatility σ. This functional satisfies the additional conditions: (Positivity) $\mathcal{F}_1(\mathbb{P}|\mathbb{P}^0) \geq 0$ and (Monotonicity) if T is an arbitrary transition probability, then

$$\mathcal{F}_1(\mathbb{P}|\mathbb{P}^0) \geq \mathcal{F}_1(T\mathbb{P}|T\mathbb{P}^0)$$

By (convex) duality, we have

PROPOSITION 2.2
The density $\mathbb{P}^*_{c_1,\ldots,c_n}(dx) = p^*(x)dx$ in (2.16) is given by

$$\frac{p^*(x)}{p^0(x)} = \left(f'^{-1}\left(\lambda^*.u(x) \right) \right)^+ \tag{2.18}$$

where

$$\lambda^* = \operatorname{argmax}_{\lambda \in \mathbb{R}^{n+2}} \lambda.U - \mathbb{E}^{\mathbb{P}^0}[\lambda.u(S_T) \left(f'^{-1}\left(\lambda.u(S_T) \right) \right)^+]$$
$$+ \mathbb{E}^{\mathbb{P}^0}[f\left(\left(f'^{-1}\left(\lambda.u(S_T) \right) \right)^+ \right)] \tag{2.19}$$

with $u(x) = \{1, x, (x - K_1)_+, \ldots, (x - K_n)_+\}$ and $U = \{1, S_0, c_1, \ldots, c_n\}$.

PROOF By introducing Lagrange's multipliers $\lambda \in \mathbb{R}^{n+2}$, I can be written as

$$I = \inf_{\mathbb{P} \in \mathcal{M}_+} \mathcal{F}(\mathbb{P}) + \sup_{\lambda \in \mathbb{R}^{n+2}} \lambda.U - \mathbb{E}^{\mathbb{P}}[\lambda.u(S_T)]$$

where $\mathcal{M}_+$ denotes the space of positive measure. From a minimax argument (see Remark 1.2),

$$I = \sup_{\lambda \in \mathbb{R}^{n+2}} \lambda.U - \sup_{p(\cdot)>0} \left(\int \lambda.u(x)p(x)dx - \int f\left(\frac{p(x)}{p^0(x)} \right) p^0(x)dx \right)$$

By taking the sup over $p > 0$, we obtain (2.18). Finally,

$$I = \sup_{\lambda \in \mathbb{R}^{n+2}} \lambda.U - \int \lambda.u(x)p^*(x)dx + \int p^0(x)f\left(\frac{p^*(x)}{p^0(x)} \right) dx$$

The supremum over λ is given by (2.19). □

The computation of p^* boils down to solving a convex optimization problem (2.19) in $\mathbb{R}^{n+2}$. This can be achieved efficiently with a Newton algorithm. In Table 2.1, we have listed some examples of functions $f(p/p^0)$ and corresponding non-parametric densities p^*. The Kullback–Leibler distance, $f(t) = t \ln t$, has been first considered in this context in [7].

Table 2.1: Examples of non-parametric density.

Chi^2 - $f(t) = (1-t)^2/2$	$p^*(x) = p^0(x)\left(1 + \lambda^*.u(x)\right)_+$
Kullback–Leibler - $f(t) = t \ln t$	$p^*(x) = p^0(x)\dfrac{e^{\lambda^*.u(x)}}{\mathbb{E}^{\mathbb{P}^0}[e^{\lambda^*.u(S_T)}]}$
$(\alpha > 0)$-divergence - $f(t) = \frac{1}{1+\alpha}\left(t^{1+\alpha} - 1\right)$	$p^*(x) = p^0(x)\left(\lambda^*.u(x)\right)_+^{\frac{1}{\alpha}}$

We illustrate our construction of marginals. We start with some liquid strikes $[0.5, 0.6, \ldots, 1.5] \times S_0$ on the 5 year EuroStock implied volatility (start date=6-Sept-11). We add an additional (moment) constraint given by the log-swap (see Section 3.2.1 for an explanation why prices for this payoff are quoted on the market)

$$\text{VS} = -\frac{2}{T}\mathbb{E}^{\mathbb{P}}[\ln \frac{S_T}{S_0}]$$

Our T-marginal should belong to the infinite-dimensional convex set

$$\mathcal{M}_{\text{VS}} = \{\mathbb{P} : \mathbb{E}^{\mathbb{P}}[S_T] = S_0, \quad \mathbb{E}^{\mathbb{P}}[(S_T - K_i)^+] = C(K_i) \equiv c_i, \quad i = 1, \ldots n,$$

$$\text{VS} = -\frac{2}{T}\mathbb{E}^{\mathbb{P}}[\ln \frac{S_T}{S_0}]\}$$

We have used two sets $\sqrt{\text{VS}} = 30.58\%$ and $\sqrt{\text{VS}} = 28.58\%$ (plots denoted "SmileCal") and the KL distance. The prior distribution $\mathbb{P}^0$ is a log-normal distribution with volatility equal to $\sqrt{\text{VS}}$. The resulting implied volatility is compared with that one used by the market (plot denoted "SmileRxPrice"), more precisely the one used at Société Générale. They are plotted in Figure 2.1. In Figure 2.2, we have plotted the densities p^*.

2.1.7   Some symmetries

2.1.7.1   Spence–Mirrlees condition

The Spence–Mirrlees condition, i.e., $c_{12} > 0$, required for the Fréchet–Hoeffding solution to hold, is very natural from a financial point of view. If we shift the payoff c by some European payoffs $\Lambda_1 \in L^1(\mathbb{P}^1)$, $\Lambda_2 \in L^1(\mathbb{P}^2)$:

$$\bar{c}(s_1, s_2) = c(s_1, s_2) + \Lambda_1(s_1) + \Lambda_2(s_2)$$

then the Monge–Kantorovich bound for $\bar{c}$ should be

$$\text{MK}_2(\bar{c}) = \text{MK}_2(c) + \mathbb{E}^{\mathbb{P}^1}[\Lambda_1(S_1)] + \mathbb{E}^{\mathbb{P}^2}[\Lambda_2(S_2)]$$

as the market price of $\Lambda_i(s_i)$ is fixed by $\mathbb{E}^{\mathbb{P}^i}[\Lambda_i(S_i)]$. The payoff $\bar{c}$ is precisely invariant under the Spence–Mirrlees condition : $\bar{c}_{12} = c_{12}$.

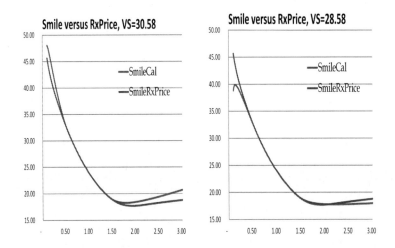

Figure 2.1: Liquid strikes $[0.5, 0.6, \ldots, 1.5] \times S_0$, STOX5E-X, 6-Sept-11. Maturity=5 years. Implied volatilities $\sigma_{\mathrm{BS}}(K)$ as a function of K with different VS volatilities: $\sqrt{\mathrm{VS}} = 30.58\%$ (left) and $\sqrt{\mathrm{VS}} = 28.58\%$ (right). The prior distribution $\mathbb{P}^0$ is a log-normal distribution with volatility equal to $\sqrt{\mathrm{VS}}$. By construction, the two curves ("SmileRxPrice" and "SmileCal") coincide at the strikes $[0.5, 0.6, \ldots, 1.5] \times S_0$. The two sets $\sqrt{\mathrm{VS}} = \{30.58\%, 28.58\%\}$ generate for the curves "SmileCal" different extrapolations for large and small strikes.

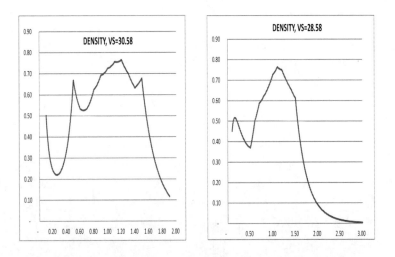

Figure 2.2: Densities with different VS volatilities $\sqrt{\mathrm{VS}} = 30.58\%$ (left) and $\sqrt{\mathrm{VS}} = 28.58\%$ (right). By construction, the densities are continuous but are not smooth - see formula p^* for the KL distance in Table 2.1.

2.1.7.2 Mirror coupling: Co-monotone rearrangement map

Similarly, the upper bound under the condition $c_{12} < 0$ is attained by the co-monotone rearrangement map

$$T(s_1) = F_2^{-1} \circ (1 - F_1(-s_1))$$

This can be obtained by applying the parity transformation $\mathcal{P}(s_1, s_2) = (-s_1, s_2)$. For each measure $\mathbb{P}$ matching the marginals $\mathbb{P}^1$ and $\mathbb{P}^2$, we can associate the measure $\mathcal{P}_*\mathbb{P}$ matching the marginals $\mathcal{P}_*\mathbb{P}^1$ and $\mathbb{P}^2$ with cumulative distributions $\bar{F}_1(s_1) \equiv 1 - F_1(-s_1)$ and $F_2(s_2)$. We conclude as the Monge–Kantorovich bounds for c and $\tilde{c}(s_1, s_2) \equiv c(-s_1, s_2)$ coincides as $\mathbb{E}^{\mathbb{P}}[c] = \mathbb{E}^{\mathcal{P}_*\mathbb{P}}[\tilde{c}]$. Similarly, by replacing c by $-c$, we obtain that the co-monotone rearrangement map gives the *lower* bound under the condition $c_{12} > 0$.

Example 2.4 Lower bound, $c(s_1, s_2) = (s_1 - K_1)^+ 1_{s_2 > K_2}$
By applying Anti-Fréchet–Hoeffding solution, the lower bound is attained by

$$\underline{\mathrm{MK}}_2 = \int_{F_2(K_2)}^{\max(1 - F_1(K_1), F_2(K_2))} (F_1^{-1}(1 - u) - K_1) du$$

with

$$\lambda_2(x) = \left(\bar{F}_1^{-1} \circ F_2(K_2) - K_1\right)^+ 1_{x > K_2}$$

$$\lambda_1(x) = (x - K_1)^+ 1_{F_2^{-1} \circ \bar{F}_1(x) > K_2} - \left(\bar{F}_1^{-1} \circ F_2(K_2) - K_1\right)^+ 1_{F_2^{-1} \circ \bar{F}_1(x) > K_2}$$

☐

Example 2.5 Lower/upper bound, $c(s_1, s_2) = (s_2/s_1 - K)^+$
For all $\mathbb{P} \in \mathcal{P}(\mathbb{P}^1, \mathbb{P}^2)$, we have

$$\int_0^1 \left(\frac{F_2^{-1}(u)}{F_1^{-1}(u)} - K\right)^+ du \leq \mathbb{E}^{\mathbb{P}}[(\frac{S_2}{S_1} - K)^+] \leq \int_0^1 \left(\frac{F_2^{-1}(1 - u)}{F_1^{-1}(u)} - K\right)^+ du$$

The bounds are attained by (Anti)-Fréchet–Hoeffding solutions. ☐

2.1.8 Robust quantile hedging

We set $p \in (0, 1]$ and define the robust quantile hedging price as

DEFINITION 2.3

$$\mathrm{MK}_p = \inf_{\mathcal{Q}^*(\mathbb{P}^1, \mathbb{P}^2)} \mathbb{E}^{\mathbb{P}^1}[\lambda_1(S_1)] + \mathbb{E}^{\mathbb{P}^2}[\lambda_2(S_2)]$$

where $Q^*(\mathbb{P}^1, \mathbb{P}^2)$ *is the set of all continuous functions* $(\lambda_1, \lambda_2) \in L^1(\mathbb{P}^1) \times L^1(\mathbb{P}^2)$ *such that for all* $\mathbb{P} \in \mathcal{P}(\mathbb{R}_+^2)$,

$$\mathbb{P}[\lambda_1(S_1) + \lambda_2(S_2) \geq c(S_1, S_2)] \geq p \qquad (2.20)$$

and $\lambda_1(s_1) + \lambda_2(s_2) \geq 0, \quad \forall\, (s_1, s_2) \in \mathbb{R}_+^2.$

Following closely the proof of Theorem 2.1, MK_p can be dualized into (see also the proof of Theorem 2.12)

$$\mathrm{MK}_p = \inf_{A \in \mathcal{B}(\mathbb{R}_+^2)\,:\, \inf_{\mathbb{P} \in \mathcal{P}(\mathbb{R}_+^2)} \mathbb{P}[A] \geq p} \sup_{\mathbb{P} \in \mathcal{P}(\mathbb{P}^1, \mathbb{P}^2)} \mathbb{E}^{\mathbb{P}}[c(S_1, S_2) 1_A]$$

Note that, $\inf_{\mathbb{P} \in \mathcal{P}(\mathbb{R}_+^2)} \mathbb{P}[A] = 0$ except if $A = \Omega$ for which $\inf_{\mathbb{P} \in \mathcal{P}(\mathbb{R}_+^2)} \mathbb{P}[A] = 1$. Therefore for all $p > 0$, $\mathrm{MK}_p = \mathrm{MK}_1$. The robust quantile hedging price coincides therefore with our super-replication price.

REMARK 2.3   Notice that Equation (2.20) should be valid for all $\mathbb{P} \in \mathcal{P}(\mathbb{R}_+^2)$ and not $\mathbb{P} \in \mathcal{P}(\mathbb{P}^1, \mathbb{P}^2)$. Indeed the market fluctuations of S_1 and S_2 are not necessarily consistent with the marginals $\mathbb{P}^1$ and $\mathbb{P}^2$. If for example we compute the price of a Vanilla on S_1 using some historical probability $\mathbb{P}^{\mathrm{hist}}$ estimated from the market, we will have in general

$$\mathbb{E}^{\mathbb{P}^{\mathrm{hist}}}[(S_1 - K)^+] \neq \mathbb{E}^{\mathbb{P}^1}[(S_1 - K)^+]$$

▯

2.1.9 Multi-marginals and infinitely-many marginals case

Most of the literature on OT focuses on the 2-asset case with a payoff $c(s_1, s_2)$. For applications in mathematical finance, it is interesting to study the case of a multi-asset payoff $c(s_1, \ldots, s_n)$ depending on n assets evaluated at the same maturity. We define the n-asset optimal transport problem (by duality) as

$$\mathrm{MK}_n \equiv \sup_{\mathbb{P} \in \mathcal{P}(\mathbb{P}^1, \ldots, \mathbb{P}^n)} \mathbb{E}^{\mathbb{P}}[c(S_1, \ldots, S_n)]$$

with $\mathcal{P}(\mathbb{P}^1, \ldots, \mathbb{P}^n) = \{\mathbb{P}\,:\, S_i \overset{\mathbb{P}}{\sim} \mathbb{P}^i,\, \forall\, i = 1, \ldots, n\}$. This problem has been studied by Gangbo [78] and recently by Carlier [36] (see also Pass [124]) with the following requirement on the payoff:

DEFINITION 2.4 see [36] $c \in C^2$ *is strictly monotone of order 2 if for all* $(i, j) \in \{1, \ldots, n\}^2$ *with* $i \neq j$, *all second order derivatives* $\partial_{ij} c$ *are positive.*

We have

THEOREM 2.4 see [78, 36]

If c is strictly monotone of order 2, there exists a unique optimal transference plan for the MK_n transport problem, and it has the form

$$\mathbb{P}^*(ds_1, \ldots, ds_n) = \mathbb{P}^1(ds_1) \prod_{i=2}^{n} \delta_{T_i(s_1)}(ds_i), \quad T_i(s) = F_i^{-1} \circ F_1(s), i = 2, \ldots, n$$

The optimal upper bound is

$$MK_n = \int c(x, T_2(x), \ldots, T_n(x)) \mathbb{P}^1(dx)$$

An extension to the infinite many marginals case has been obtained recently by Pass [125] who studies

$$MK_\infty \equiv \sup_{\mathbb{P} \,:\, S_t \overset{\mathbb{P}}{\sim} \mathbb{P}^t, \,\forall\, t \in (0,T]} \mathbb{E}[h\left(\int_0^T S_t dt\right)]$$

where h is a convex function. Let F_t the cumulative distribution of $\mathbb{P}^t$. Define the stochastic process $S_t^{\text{opt}}(\omega) = F_t^{-1}(\omega)$, $\omega \in [0,1]$. The underlying probability space is the interval $[0,1]$ with Lebesgue measure. We have

THEOREM 2.5 see [125]

The process $S_.^{\text{opt}}$ is the unique maximizer in MK_∞.

2.1.10 Link with Hamilton–Jacobi equation

Here we take a cost function $c(s_1, s_2) = L(s_2 - s_1)$ with L a strictly concave function such that the Spence–Mirrlees condition is satisfied. From the formulation (2.8), one can link the Monge–Kantorovich formulation to the solution of a Hamilton–Jacobi equation through the Hopf–Lax formula:

PROPOSITION 2.3 see e.g. [139]

$$MK_2 = \inf_{u(1,\cdot)} -\mathbb{E}^{\mathbb{P}^1}[u(0, S_1)] + \mathbb{E}^{\mathbb{P}^2}[u(1, S_2)]$$

where $u(0, \cdot)$ is the (viscosity) solution at $t = 0$ of the following HJ equation with terminal boundary condition $u(1, \cdot)$:

$$\partial_t u + H(Du) = 0, \quad H(p) \equiv \inf_q \{pq - L(q)\} \tag{2.21}$$

PROOF From the dynamic programming principle, u, satisfying Hamilton–Jacobi equation (2.21), can be written as

$$u(t,x) = \inf_{\zeta} u\left(1, x + \int_t^1 \dot{\zeta}(s)ds\right) - \int_t^1 L(\dot{\zeta}(s))ds$$

The minimisation over $\dot{\zeta}$ gives that $\dot{\zeta}$ is a constant q (Fréchet derivative with respect to $\dot{\zeta}$ gives the critical equation $\frac{d^2\zeta(t)}{dt^2} = 0$).

$$u(t,x) = \inf_q u(1, x + q(1-t)) - L(q)(1-t)$$

By setting $y = x + q(1-t)$, we get Hopf–Lax's formula:

$$u(t,x) = \inf_y u(1,y) - L\left(\frac{y-x}{1-t}\right)(1-t)$$

For $t = 0$, this gives that $-u(0, \cdot)$ is the L-transform of $u(1, \cdot)$: $-u(0,x) = \sup_y L(y-x) - u(1,y)$. We conclude with Proposition 2.1. □

In the next section, we introduce a martingale version of OT, first developed in [17, 77] where we have obtained a Monge–Kantorovich duality result.

2.2 Martingale optimal transport

We consider a payoff $c(s_1, s_2)$ depending on a *single* asset evaluated at two dates $t_1 < t_2$. As above, we assume that Vanilla options of all strikes with maturities t_1 and t_2 are traded and therefore, one can imply the distribution of S_1 and S_2 (here $S_1 \equiv S_{t_1}, S_2 \equiv S_{t_2}$). The model-independent upper bound, consistent with t_1 and t_2 Vanilla options, can then be defined as a martingale version of the above OT problem MK_2:

DEFINITION 2.5

$$\widetilde{\mathrm{MK}}_2 \equiv \inf_{\mathcal{M}^*(\mathbb{P}^1, \mathbb{P}^2)} \mathbb{E}^{\mathbb{P}^1}[\lambda_1(S_1)] + \mathbb{E}^{\mathbb{P}^2}[\lambda_2(S_2)] \tag{2.22}$$

where $\mathcal{M}^*(\mathbb{P}^1, \mathbb{P}^2)$ is the set of functions $\lambda_1 \in \mathrm{L}^1(\mathbb{P}^1), \lambda_2 \in \mathrm{L}^1(\mathbb{P}^2)$ and H a bounded continuous function on $\mathbb{R}_+$ such that

$$\lambda_1(s_1) + \lambda_2(s_2) + H(s_1)(s_2 - s_1) \geq c(s_1, s_2), \quad \forall (s_1, s_2) \in \mathbb{R}_+^2 \tag{2.23}$$

This corresponds to a semi-static hedging strategy which consists in holding European payoffs λ_1 and λ_2 and applying a delta strategy at t_1, generating a

P&L $H(s_1)(s_2-s_1)$ at t_2 with zero cost. We could also add a term $H_0(S_0)(s_1-S_0)$ corresponding to performing a delta-hedging at $t = 0$. As this term can be incorporated into $\lambda_1(s_1)$, it is not included. Similarly, an intermediate delta-hedging term $H_i(S_0, \ldots, s_{t_i})(s_{t_{i+1}} - s_{t_i})$ where $0 < t_i < t_{i+1} \le t_2$ can be added but it can be shown that the optimal solution is attained for $H_i = 0$. These terms are therefore not needed and will be disregarded next (see Corollary 2.1).

Note that in comparison with the OT MK_2 previously reported, we have $\widetilde{\mathrm{MK}}_2 \le \mathrm{MK}_2$ due to the appearance of the function H.

At this point, a natural question is how the classical results in OT generalize in the present martingale version. We follow closely our introduction of OT and explain how the various concepts previously explained extend to the present setting. Our research partly originates from a systematic derivation of Skorokhod embedding solutions and understanding of particle methods for non-linear McKean stochastic differential equations appearing in the calibration of financial models (see Section 4.2.4). From a practical point of view, the derivation of these optimal bounds allows to better understand the risk of exotic options as illustrated in the next example.

Example 2.6 Forward-start options, see [93]

We consider model-independent bounds for forward-start options with payoff $(s_2/s_1 - K)^+$. EuroStock Vanillas (pricing date = 2 Feb. 2010) at time $t_1 = 1$ year and $t_2 = 1.5$ years are quoted on the market with $N = 18$ strikes ranging in $[30\%, 200\%] \times S_0$. $\widetilde{\mathrm{MK}}_2$ (i.e., upper bound) reads as (with $K_1^0 = K_2^0 = 0$ corresponding to forward prices: $\mathbb{E}^{\mathbb{P}^1}[S_1] = \mathbb{E}^{\mathbb{P}^2}[S_2] = S_0 = C(t_1, K_1^0) = C(t_2, K_2^0)$)

$$U(K) \equiv \min_{\nu, (\omega_1^j), (\omega_2^j), H(\cdot)} \nu + \sum_{j=0}^{N} \omega_1^j C(t_1, K_1^j) + \sum_{j=0}^{N} \omega_2^j C(t_2, K_2^j)$$

$$F(s_1, s_2) \ge \left(\frac{s_2}{s_1} - K\right)^+, \quad \forall (s_1, s_2) \in (\mathbb{R}_+)^2$$

with $F \equiv \sum_{j=0}^{N} \omega_1^j (s_1 - K_j)^+ + \sum_{j=0}^{N} \omega_2^j (s_2 - K_j)^+ + \nu + H(s_1)(s_2 - s_1)$ and $C(t_i, K)$ the market value of a call of maturity t_i and strike K. The linear program $U(K)$ can be solved numerically using a simplex algorithm by discretizing s_1 and s_2 on a two-dimensional grid. The lower bound can be obtained similarly by replacing min by max and $F(s_1, s_2) \ge \left(\frac{s_2}{s_1} - K\right)^+$ by $F(s_1, s_2) \le \left(\frac{s_2}{s_1} - K\right)^+$. We have compared the upper and lower bounds (for different strikes K) against prices produced by various (stochastic) volatility models commonly used by practitioners (see Figure 2.3): Dupire Local volatility model (in short LV) [68], Bergomi model [23, 24] which is a two-factor variance curve model (see Remark 2.4) and the local Bergomi model

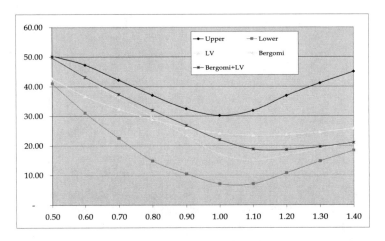

Figure 2.3: Implied volatility for cliquet option (defined as the volatility σ, which when put in the Black–Scholes formula with $S_0 = 1$ and $T = t_2 - t_1$ reproduces the price of cliquet option) as a function of K. Lower/Upper bounds versus (local) Bergomi and LV models for cliquet options (quoted in Black–Scholes volatility×100). Parameters for the Bergomi model: $\sigma = 2.0$, $\theta = 22.65\%$, $k_1 = 4$, $k_2 = 0.125$, $\rho = 34.55\%$, $\rho_{\text{SX}} = -76.84\%$, $\rho_{\text{SY}} = -86.40\%$.

(in short BergomiLV) [91, 87] which has the property to be calibrated to the implied volatility surface.

LV and BergomiLV have been calibrated to the EuroStock implied volatility (pricing date = 2 Feb. 2010). The Bergomi model has been calibrated to the variance-swap term structure $T \mapsto -\frac{2}{T}\mathbb{E}^{\mathbb{P}^{\text{mkt}}}[\ln \frac{S_T}{S_0}]$.

We should emphasize that lower and upper bounds for each strike K correspond to a different martingale measure (as $\left(\frac{s_2}{s_1} - K\right)^+$ does not satisfy the condition $c_{122} > 0$, see Theorem 2.8). This is not the case if we do not include the martingale constraint as the upper/lower bounds are attained by the Fréchet–Hoeffding bounds for each strike (here $-c_{12} > 0$ holds) - see Example 2.5.

For $K = 1$, the difference between the implied volatility given by the upper bound and the Bergomi model is around 10%. This result shows that forward-start options are poorly constrained by the Vanilla implied volatility. As a conclusion, the (common) practice to calibrate stochastic volatility model on Vanillas to price exotic options (depending strongly on forward/skew volatility) is inappropriate: forward-start options are poorly hedged by t_1 and t_2 Vanilla options. □

REMARK 2.4 Two-factor Bergomi model For completeness, as this model will be used in our numerical experiments, the dynamics for the two-factor Bergomi model reads

$$dSt = S_t \sqrt{\xi_t^t} dW_t$$
$$\xi_t^T = \xi_0^T f^T(t, x_t^T), \quad f^T(t, x) = \exp(2\sigma x - 2\sigma^2 h(t, T))$$
$$x_t^T = \alpha_\theta \left((1 - \theta) e^{-k_1(T-t)} X_t + \theta e^{-k_2(T-t)} Y_t \right)$$
$$\alpha_\theta = \left((1 - \theta)^2 + \theta^2 + 2\rho\theta(1 - \theta) \right)^{-1/2}$$
$$dX_t = -k_1 X_t dt + dW_t^X, \quad dY_t = -k_2 Y_t dt + dW_t^Y$$
$$d\langle W^X, W^Y \rangle_t = \rho dt, \quad d\langle W, W^X \rangle_t = \rho_{SX} dt, \quad d\langle W, W^Y \rangle_t = \rho_{SY} dt$$

where

$$h(t, T) = (1 - \theta)^2 e^{-2k_1(T-t)} \mathbb{E}\left[X_t^2 \right] + \theta^2 e^{-2k_2(T-t)} \mathbb{E}\left[Y_t^2 \right]$$
$$+ 2\theta(1 - \theta) e^{-(k_1+k_2)(T-t)} \mathbb{E}\left[X_t Y_t \right]$$
$$\mathbb{E}\left[X_t^2 \right] = \frac{1 - e^{-2k_1 t}}{2k_1}, \quad \mathbb{E}\left[Y_t^2 \right] = \frac{1 - e^{-2k_2 t}}{2k_2}$$
$$\mathbb{E}\left[X_t Y_t \right] = \rho \frac{1 - e^{-(k_1+k_2)t}}{k_1 + k_2}$$

This model, commonly used by practitioners, is a variance swap curve model that admits a two-dimensional (three-dimensional if we consider time as one of the Markov processes) Markovian representation.

In the local Bergomi model, we have

$$dS_t = S_t \sigma(t, S_t) \sqrt{\xi_t^t} dW_t$$

where $\sigma(\cdot, \cdot)$ is chosen for matching the market price of Vanillas (at $t = 0$) - See Section 4.2.4. □

2.2.1 Dual formulation

In this section, we establish a dual version similar to the Monge–Kantorovich formulation. We set

$$\widetilde{\mathrm{MK}}_2^* \equiv \sup_{\mathbb{P} \in \mathcal{M}(\mathbb{P}^1, \mathbb{P}^2)} \mathbb{E}^{\mathbb{P}}[c(S_1, S_2)] \tag{2.24}$$

where $\mathcal{M}(\mathbb{P}^1, \mathbb{P}^2) = \{\mathbb{P} : \mathbb{E}^{\mathbb{P}}[S_2|S_1] = S_1, S_1 \overset{\mathbb{P}}{\sim} \mathbb{P}^1, S_2 \overset{\mathbb{P}}{\sim} \mathbb{P}^2\}$ denotes the set of (discrete) martingale measures on $\mathbb{R}_+^2$ with marginals $\mathbb{P}^1$ and $\mathbb{P}^2$.

2.2.1.1 Set $\mathcal{M}(\mathbb{P}^1, \mathbb{P}^2)$: some properties

As $\mathcal{M}(\mathbb{P}^1, \mathbb{P}^2)$ is convex and weakly compact[5], the dual $\widetilde{\mathrm{MK}}_2^*$ is attained by an extremal point. The following proposition gives a necessary and sufficient condition to ensure that $\mathcal{M}(\mathbb{P}^1, \mathbb{P}^2)$ is non-empty:

> **PROPOSITION 2.4 see Kellerer, [111], and also Corollary 4.2**
> *The set $\mathcal{M}(\mathbb{P}^1, \mathbb{P}^2)$ is feasible (i.e., non-empty) if and only if $\mathbb{P}^1, \mathbb{P}^2$ have the same mean S_0 and $\mathbb{P}^1 \leq \mathbb{P}^2$ are in convex order.*

Convex order means that

> **DEFINITION 2.6 Convex order** $\quad \mathbb{P}^1 \leq \mathbb{P}^2$ *are in convex order if and only if*
>
> $$\mathbb{E}^{\mathbb{P}^1}[(S_1 - K)^+] \leq \mathbb{E}^{\mathbb{P}^2}[(S_2 - K)^+], \quad \forall K \in \mathbb{R}_+$$

In financial term, we exclude calendar spread arbitrage opportunities (on call options). The necessary condition can easily be obtained by applying the Jensen inequality on $(S_2 - K)^+$ as $\mathbb{E}[S_2 | S_1] = S_1$: For all $\mathbb{P} \in \mathcal{M}(\mathbb{P}^1, \mathbb{P}^2)$,

$$\mathbb{E}^{\mathbb{P}^2}[(S_2 - K)^+] = \mathbb{E}^{\mathbb{P}}[(S_2 - K)^+] \geq \mathbb{E}^{\mathbb{P}}[(S_1 - K)^+] = \mathbb{E}^{\mathbb{P}^1}[(S_1 - K)^+]$$

The sufficient condition can be deduced by explicitly building a martingale measure matching marginals $\mathbb{P}^1$ and $\mathbb{P}^2$ from the Dupire local volatility model (see e.g. [99] and Section 4.2.6).

Extremal points

From the Krein–Milman theorem, $\mathcal{M}(\mathbb{P}^1, \mathbb{P}^2)$ can be reconstructed from its extremal points:

$$\mathcal{M}(\mathbb{P}^1, \mathbb{P}^2) = \overline{\mathrm{Conv}(\mathrm{Ext}(\mathcal{M}(\mathbb{P}^1, \mathbb{P}^2)))}$$

The following theorem gives a characterization of the extremal points:

> **THEOREM 2.6 [35]**
> *Let $\mathbb{P} \in \mathcal{M}(\mathbb{P}^1, \mathbb{P}^2)$. The following two properties are equivalent:*
>
> *1. $\mathbb{P} \in \mathrm{Ext}(\mathcal{M}(\mathbb{P}^1, \mathbb{P}^2))$.*

---

[5]We have already seen that $\mathcal{P}(\mathbb{P}^1, \mathbb{P}^2) \supset \mathcal{M}(\mathbb{P}^1, \mathbb{P}^2)$ is relatively compact. We conclude by showing that $\mathcal{M}(\mathbb{P}^1, \mathbb{P}^2)$ is closed for the weak topology, see [17].

2. *The set of all functions $c \in L^1(\mathbb{P})$ that can be represented as*

$$c(s_1, s_2) = \lambda_1(s_1) + \lambda_2(s_2) + H_1(s_1)(s_2 - s_1), \quad \mathbb{P} - a.s.$$

for some $\lambda_1 \in L^1(\mathbb{P}^1)$, $\lambda_2 \in L^2(\mathbb{P}^2)$, and $H \in L^0(\mathbb{P}^1)$, is dense in $L^1(\mathbb{P})$.

This theorem shows that a payoff $c \in L^1(\mathbb{P})$ can be (approximately) semi-statically replicated and therefore an element $\mathbb{P} \in \text{Ext}(\mathcal{M}(\mathbb{P}^1, \mathbb{P}^2))$ corresponds to a robust complete model (i.e., all payoffs $c \in L^1(\mathbb{P})$ are attainable by a semi-static hedge).

2.2.1.2 Duality

THEOREM 2.7 see [17]
Assume that $\mathbb{P}^1 \leq \mathbb{P}^2$ are probability measures on $\mathbb{R}_+$ with first moments S_0 and Assumption 3 holds. Then there is no duality gap, i.e., $\widetilde{\text{MK}}_2^* = \widetilde{\text{MK}}_2$. Moreover, the dual value $\widetilde{\text{MK}}_2^*$ is attained, i.e., there exists a martingale measure $\mathbb{P}^* \in \mathcal{M}(\mathbb{P}^1, \mathbb{P}^2)$ such that $\widetilde{\text{MK}}_2^* = \mathbb{E}^{\mathbb{P}^*}[c]$. In general, the primal $\widetilde{\text{MK}}_2$ is not attained (see [17] for a counterexample).

Note that in [20] a quasi-sure formulation of the dual problem is introduced and shown to yield a similar duality with existence of primal optimizers.

PROOF (Sketch) The function H is interpreted as the Lagrange multiplier associated to the martingale condition. Then, the proof copycats the one in Theorem 2.1. □

Note that the quadratic cost $c(s_1, s_2) = (s_2 - s_1)^2$, which is the main example in OT, is here degenerate as

$$\mathbb{E}^{\mathbb{P}}[(S_2 - S_1)^2] = \mathbb{E}^{\mathbb{P}^2}[S_2^2] - \mathbb{E}^{\mathbb{P}^1}[S_1^2], \quad \mathbb{P} \in \mathcal{M}(\mathbb{P}^1, \mathbb{P}^2)$$

meaning that the cost value is identical for all $\mathbb{P} \in \mathcal{M}(\mathbb{P}^1, \mathbb{P}^2)$.
From our duality result, we derive the following corollary that shows that delta hedging at intermediate dates are not needed:

COROLLARY 2.1
Define

$$\widetilde{\text{MK}}_2^{3/2} \equiv \inf_{\mathcal{M}_{3/2}^*(\mathbb{P}^1, \mathbb{P}^2)} \mathbb{E}^{\mathbb{P}^1}[\lambda_1(S_1)] + \mathbb{E}^{\mathbb{P}^2}[\lambda_2(S_2)] \tag{2.25}$$

where $\mathcal{M}_{3/2}^*(\mathbb{P}^1, \mathbb{P}^2)$ is the set of functions $\lambda_1 \in L^1(\mathbb{P}^1), \lambda_2 \in L^1(\mathbb{P}^2)$ and H (resp. $H_{3/2}$) bounded continuous function on $\mathbb{R}_+$ (resp. $\mathbb{R}_+^2$) such that

$\forall (s_1, s_{3/2}, s_2) \in \mathbb{R}_+^3,$

$$\lambda_1(s_1) + \lambda_2(s_2) + H(s_1)(s_{\frac{3}{2}} - s_1) + H_{3/2}(s_1, s_{3/2})(s_2 - s_{3/2}) \geq c(s_1, s_2)$$

Then,

$$\widetilde{\mathrm{MK}}_2^{3/2} = \widetilde{\mathrm{MK}}_2$$

PROOF   By duality, we have

$$\widetilde{\mathrm{MK}}_2^{3/2} \overset{\text{Duality}}{=} \sup_{\mathcal{M}_{3/2}(\mathbb{P}^1, \mathbb{P}^2)} \mathbb{E}^{\mathbb{P}}[c(S_1, S_2)]$$

where $\mathcal{M}_{3/2}(\mathbb{P}^1, \mathbb{P}^2) = \{\mathbb{P} : \mathbb{E}^{\mathbb{P}}[S_2 | S_{3/2}, S_1] = S_{3/2}, \mathbb{E}^{\mathbb{P}}[S_{3/2} | S_1] = S_1, S_1 \overset{\mathbb{P}}{\sim} \mathbb{P}^1, S_2 \overset{\mathbb{P}}{\sim} \mathbb{P}^2\}$ denotes the set of (discrete) martingale measures on $\mathbb{R}_+^3$ with marginals $\mathbb{P}^1$ and $\mathbb{P}^2$ at t_1 and t_2. As $\mathcal{M}^*(\mathbb{P}^1, \mathbb{P}^2) \subset \mathcal{M}_{3/2}^*(\mathbb{P}^1, \mathbb{P}^2)$ (take $H_{3/2} = H$), this implies

$$\widetilde{\mathrm{MK}}_2^{3/2} \leq \widetilde{\mathrm{MK}}_2 = \mathbb{E}^{\mathbb{P}^*}[c(S_1, S_2)], \quad \mathbb{P}^* \in \mathcal{M}(\mathbb{P}^1, \mathbb{P}^2)$$

where we have used that $\widetilde{\mathrm{MK}}_2$ is attained by $\mathbb{P}^*$ from weak compactness. If we define $\mathbb{P}^{*,3/2}(ds_1, ds_{3/2}, ds_2) = \mathbb{P}^*(ds_{3/2}, ds_2)\delta(s_{3/2} - s_1)ds_1$, then $\mathbb{P}^{*,3/2} \in \mathcal{M}_{3/2}(\mathbb{P}^1, \mathbb{P}^2)$ and

$$\mathbb{E}^{\mathbb{P}^{*,3/2}}[c(S_1, S_2)] \leq \widetilde{\mathrm{MK}}_2^{3/2}$$

This implies our result as $\mathbb{E}^{\mathbb{P}^{*,3/2}}[c(S_1, S_2)] = \mathbb{E}^{\mathbb{P}^*}[c(S_1, S_2)]$.   □

REMARK 2.5 Duality gap   In [17], we assume a weaker assumption on the payoff: c can be an upper semicontinuous function and Assumption 3 holds. The upper semicontinuity requirement on c can however not be relaxed as observed by the following example in [20]. We consider a bounded lower semicontinuous payoff $c(s_1, s_2) = 1_{s_1 \neq s_2}$ with $\mathbb{P}^1 = \mathbb{P}^2 = U([0, 1])$ the uniform distribution on $[0, 1]$. Then there is a duality gap as $\widetilde{\mathrm{MK}}_2^* = 0$ which is attained by $\mathbb{P}^*(ds_1, ds_2) = \delta(s_2 - s_1)ds_1 ds_2$ the uniform distribution on the diagonal of $[0, 1]^2$ and $\widetilde{\mathrm{MK}}_2 = 1$ which is attained by $\lambda_1^* = 0, \lambda_2^* = 1$ and $H^* = 0$. Indeed, on the primal side, we should have

$$\lambda_1^*(s_1) + \lambda_2^*(s_2) + H^*(s_1)(s_2 - s_1) \geq 1_{s_1 \neq s_2}, \forall (s_1, s_2) \in [0, 1]^2$$

Taking $s_2 = s_1 + \epsilon$, we get

$$\lambda_1^*(s_1) + \lambda_2^*(s_1 + \epsilon) + \geq 1 - H^*(s_1)\epsilon, \forall s_1 \in [0, 1]$$

By taking the expectation with respect to $\mathbb{P}^1 = \mathbb{P}^2$, we obtain

$$\mathbb{E}^{\mathbb{P}^1}[\lambda_1^*(s_1)] + \mathbb{E}^{\mathbb{P}^2}[\lambda_2^*(s_2 + \epsilon)] \geq 1 - \mathbb{E}^{\mathbb{P}^1}[H^*(s_1)]\epsilon$$

Letting $\epsilon \to 0$, this implies

$$\mathbb{E}^{\mathbb{P}^1}[\lambda_1^*(s_1)] + \mathbb{E}^{\mathbb{P}^2}[\lambda_2^*(s_2)] \geq 1$$

which is attained by $\lambda_1^* = 0, \lambda_2^* = 1$ and $H^* = 0$.

□

Example 2.7 Monge's cost $c(s_1, s_2) = |s_2 - s_1|$

Our dual formulation will be a key tool to compute explicitly $\widetilde{\mathrm{MK}}_2$. This is illustrated on the following example drawn from [103]. We take $\mathbb{P}^1 = U([-1, 1])$, $\mathbb{P}^2 = U([-2, 2])$ and $c(s_1, s_2) = |s_2 - s_1|$. As an element $\mathbb{P}^* \in \mathcal{M}(\mathbb{P}^1, \mathbb{P}^2)$, we take $S_2 = S_1 + Z$ with $S_1 \sim \mathbb{P}^1$ and $\mathbb{P}[Z = \pm 1] = 1/2$. We deduce that

$$1 \leq \widetilde{\mathrm{MK}}_2^*$$

Since $|x| \leq \frac{1}{2}(x^2 + 1)$ and $(s_2 - s_1)^2 = s_2^2 - s_1^2 - 2s_1(s_2 - s_1)$, an element $(\lambda_1^*, \lambda_2^*, H^*) \in \mathcal{M}^*(\mathbb{P}^1, \mathbb{P}^2)$ is $\lambda_1^*(s_1) = \frac{1}{2} - \frac{s_1^2}{2}$, $\lambda_2^*(s_2) = \frac{s_2^2}{2}$ and $H^*(s_1) = -s_1$: For all $(s_1, s_2) \in [-1, 1] \times [-2, 2]$,

$$|s_2 - s_1| \leq \frac{1}{2} - \frac{s_1^2}{2} + \frac{s_2^2}{2} - s_1(s_2 - s_1)$$

This implies that

$$\widetilde{\mathrm{MK}}_2 \leq 1$$

The (weak) duality result $\widetilde{\mathrm{MK}}_2^* \leq \widetilde{\mathrm{MK}}_2$ implies that $\widetilde{\mathrm{MK}}_2 = \widetilde{\mathrm{MK}}_2^* = 1$. □

2.2.2 Link with Hamilton–Jacobi–Bellman equation

In OT, one can prove that the minimisation can be restricted to the class of c-concave function (see equation (2.8)). A similar result holds in MOT where the c-concave property is replaced by the concave envelope:

PROPOSITION 2.5 see [17]
Assume that $\mathbb{P}^1 \leq \mathbb{P}^2$ are probability measures on $\mathbb{R}_+$ with first moments S_0 and Assumption 3 holds. Then

$$\widetilde{\mathrm{MK}}_2 = \inf_{\lambda \in L^1(\mathbb{P}^2)} \mathbb{E}^{\mathbb{P}^1}[(c(S_1, \cdot) - \lambda(\cdot))^{**}(S_1)] + \mathbb{E}^{\mathbb{P}^2}[\lambda(S_2)] \qquad (2.26)$$

where for a function g, g** denotes its concave envelope, i.e., the smallest concave function greater than or equal to g.

PROOF   (i): We show first that the formula (2.26) is greater than or equal to the right hand side of (2.24). Let $\lambda_2 \in L^1(\mathbb{P}^2)$. For $\mathbb{P} \in \mathcal{M}(\mathbb{P}^1, \mathbb{P}^2)$ satisfying $\mathbb{E}^{\mathbb{P}}[c(S_1, S_2)] < \infty$ we have

$$\mathbb{E}^{\mathbb{P}}[c(S_1, S_2)] = \mathbb{E}^{\mathbb{P}}[c(S_1, S_2) - \lambda_2(S_2)] + \mathbb{E}^{\mathbb{P}^2}[\lambda_2(S_2)]$$

$$\leq \mathbb{E}^{\mathbb{P}}[(c(S_1, S_2) - \lambda_2(S_2))^{**}] + \mathbb{E}^{\mathbb{P}^2}[\lambda_2(S_2)]$$

$$= \mathbb{E}^{\mathbb{P}^1}[\mathbb{E}^{\mathbb{P}}[(c(S_1, S_2) - \lambda_2(S_2))^{**} | S_1]] + \mathbb{E}^{\mathbb{P}^2}[\lambda_2(S_2)] \qquad (2.27)$$

$$\leq \mathbb{E}^{\mathbb{P}^1}[(c(S_1, \mathbb{E}_{\mathbb{P}}[S_2 | S_1]) - \lambda_2(\mathbb{E}^{\mathbb{P}}[S_2 | S_1]))^{**}] + \mathbb{E}^{\mathbb{P}^2}[\lambda_2(S_2)] \qquad (2.28)$$

$$= \mathbb{E}^{\mathbb{P}^1}[(c(S_1, \cdot) - \lambda_2(\cdot))^{**}(S_1)] + \mathbb{E}^{\mathbb{P}^2}[\lambda_2(S_2)],$$

where the inequality between (2.27) and (2.28) holds due to Jensen's inequality. This proves the first inequality.

(ii): To establish the reverse inequality, we make a simple observation. Let $s_1 \in \mathbb{R}_+$ and $g \colon \mathbb{R}_+ \to \mathbb{R}$ be some function. Suppose that for $\lambda_1 \in \mathbb{R}$ there exists $H \in \mathbb{R}$ such that

$$\lambda_1 + H \cdot (s_2 - s_1) \geq g(s_2)$$

for all $s_2 \in \mathbb{R}_+$. Then $\lambda_1 \geq g^{**}(s_1)$. Applying this for $s_1 \in \mathbb{R}_+$ to the function $s_2 \mapsto g(s_2) = c(s_1, s_2) - \lambda_2(s_2)$ we obtain

$$\inf_{\lambda_2} \mathbb{E}^{\mathbb{P}^1}[(c(S_1, S_1) - \lambda_2(S_1))^{**}] + \mathbb{E}^{\mathbb{P}^2}[\lambda_2(S_2)]$$

$$\leq \inf_{\lambda_2} \sup_{\lambda_1 \,:\, \exists H, \lambda_1(s_1) + H(s_1)(s_2 - s_1) \geq c - \lambda_2(s_2)} \mathbb{E}^{\mathbb{P}^1}[\lambda_1(S_1)] + \mathbb{E}^{\mathbb{P}^2}[\lambda_2(S_2)] = \widetilde{\mathrm{MK}}_2$$

$$\square$$

From formulation (2.26), one can link the martingale Monge–Kantorovich formulation to the solution of a Hamilton–Jacobi–Bellman equation:

COROLLARY 2.2

$$\widetilde{\mathrm{MK}}_2 = \inf_{u(1, \cdot) \in L^1(\mathbb{P}^2)} \mathbb{E}^{\mathbb{P}^1}[u(0, S_1, S_1)] + \mathbb{E}^{\mathbb{P}^2}[u(1, S_2)]$$

where

$$u(0, s_1, s) \equiv \sup_{\sigma. \in [0, \infty]} \mathbb{E}^{\mathbb{P}}[c(s_1, S_T) - u(1, S_T) | S_0 = s]$$

with $dS_t = \sigma_t dB_t$. B *is a Brownian motion and σ. is an adapted (with respect to the filtration of B.) unbounded control process.*

For fixed s_1, $u(0, s_1, s)$ corresponds to the value function of a (singular) stochastic control problem which consists in maximizing the expectation of $c(s_1, S_T) - u(1, S_T)$ over all controls σ. In financial terms, the SDE for S_t corresponds to an unbounded uncertain volatility model.

PROOF $u(0, s_1, s)$ is the viscosity solution of the HJB equation ([129], see Theorem 3.5)

$$\max\left(c(s_1, s) - u(1, s) - u(0, s_1, s), \partial_{ss} u(0, s_1, s)\right) = 0$$

The solution is $u(0, s_1, \cdot) = (c(s_1, \cdot) - u(1, \cdot))^{**}$ from which we complete the proof with Proposition 2.5. □

2.2.3 A discrete martingale Fréchet–Hoeffding solution

In this section (see [95] for details), we solve explicitly $\widetilde{MK_2}$ under the martingale Spence–Mirrlees condition $\partial_{s_1 s_2 s_2} c > 0$. This gives a martingale measure, similar in spirit to the Fréchet–Hoeffding solution: under the new condition $\partial_{s_1 s_2 s_2} c > 0$, the optimal measure is payoff-independent and depends only on the marginals $\mathbb{P}^1$ and $\mathbb{P}^2$. Moreover, we will show that this condition guarantees that the primal infimum is attained which was not guaranteed from our previous duality result (see Theorem 2.7).

The optimal measure is no more supported on a single map T as it was for Fréchet–Hoeffding solution. Indeed, the martingale constraint can not be fulfilled in this case (except in the trivial case $\mathbb{P}^1 = \mathbb{P}^2$):

$$\mathbb{E}^{\mathbb{P}^*}\left[(S_2 - S_1)|S_1 = s_1\right] = T(s_1) - s_1 \neq 0$$

for

$$\mathbb{P}^*(ds_1, ds_2) = \delta_{T(s_1)}(ds_2)\mathbb{P}^1(ds_1), \quad T(x) = F_2^{-1} \circ F_1(x)$$

As a straightforward guess, we will assume that $\mathbb{P}^*$ is supported along two maps $s_2 = T_d(s_1)$ and $s_2 = T_u(s_1)$. This intuition comes from [105] where the authors consider the problem of finding the optimal upper bound on the price of a forward-start straddle $|s_2 - s_1|$ and from an old result of Dubins, Schwarz [66] that characterizes the extreme points of the space of all distributions of discrete martingales as those that possess these two properties : (a) S_0 is fixed and (b) the conditional distribution of each S_k given the past up to time $k-1$ is almost surely a two-valued distribution. This result should be put in light with the fact that a pricing binomial model is complete, and therefore a payoff can be dynamically replicated under this model. Indeed, if we take for granted that the primal is attained (by $(\lambda_1^*, \lambda_2^*, H^*)$), the duality result implies that the payoff can be perfectly dynamically replicated under $\mathbb{P}^* \in \mathcal{M}(\mathbb{P}^1, \mathbb{P}^2)$:

$$\lambda_1^*(s_1) + \lambda_2^*(s_2) + H^*(s_1)(s_2 - s_1) = c(s_1, s_2), \quad \mathbb{P}^* - \text{a.s.} \qquad (2.29)$$

for the optimal martingale measure $\mathbb{P}^*$. Indeed, $\lambda_1^*(s_1) + \lambda_2^*(s_2) + H^*(s_1)(s_2 - s_1) - c(s_1, s_2)$ is a positive r.v. with zero mean as

$$\mathbb{E}^{\mathbb{P}^*}[\lambda_1^*(S_1) + \lambda_2^*(S_2) + H^*(S_1)(S_2 - S_1) - c(S_1, S_2)] =$$
$$\mathbb{E}^{\mathbb{P}^1}[\lambda_1^*(S_1)] + \mathbb{E}^{\mathbb{P}^2}[\lambda_2^*(S_2)] - \mathbb{E}^{\mathbb{P}^*}[c(S_1, S_2)] = 0$$

Therefore Equation (2.29) should hold.

Explicit solution for the primal and dual

The extremal probability measure $\mathbb{P}^*$ was first characterized by Beiglböck–Juillet [18] who proved that $\mathbb{P}^*$ coincides with the unique left-monotone martingale transference plan defined as:

DEFINITION 2.7 *We say that $\mathbb{P} \in \mathcal{M}(\mathbb{P}^1, \mathbb{P}^2)$ is left-monotone if there exists a Borel set $\Gamma \subset \mathbb{R} \times \mathbb{R}$ such that $\mathbb{P}[(X, Y) \in \Gamma] = 1$, and for all $(x, y_1), (x, y_2), (x', y') \in \Gamma$ with $x < x'$, it must hold that $y' \notin (y_1, y_2)$.*

Below, by using our dual formulation, we give an explicit characterization of $\mathbb{P}^*$ in terms of ODEs.

Assumption 4 (Assum$(\mathbb{P}^1, \mathbb{P}^2)$) *For the sake of simplicity, we will assume that $\delta F \equiv F_2 - F_1$ has a unique maximum m. In practice, distributions implied from Vanilla market satisfy this condition (see Figure 4.1). The general case is considered in [95].*

Dual side

We define $\mathbb{P}^*$ as

$$\mathbb{P}^*(ds_1, ds_2) = \mathbb{P}^1(ds_1)\left(q(s_1)\delta_{T_u(s_1)}(ds_2) + (1 - q(s_1))\delta_{T_d(s_1)}(ds_2)\right),$$
$$q(x) = \frac{x - T_d(x)}{T_u(x) - T_d(x)} \qquad (2.30)$$

with the maps $T_d(x) \leq x \leq T_u(x)$, T_u increasing, T_d decreasing defined by

$$T_u(x) = T_d(x) = x, \quad x \leq m$$
$$T_u(x) = F_2^{-1}(F_1(x) + \delta F(T_d(x))) \qquad (2.31)$$
$$T_d'(x) = -\frac{T_u(x) - x}{T_u(x) - T_d(x)} \frac{F_1'(x)}{F_2'(T_d(x)) - F_1'(T_d(x))} \qquad (2.32)$$

Finding $T_d(x)$ for $x \geq m$ boils down solving the first-order ODE (2.32) with the initial condition $T_d(m) = m$, $T_u(x)$ being given by (2.31). An equivalent expression for $T_d(x)$ for $x \geq m$ is given as the unique solution $t \in \mathbb{R}_+$ (with $t \leq m \leq x$) of

$$t = T_d(x) \text{ s.t. } -\int_t^m (g(x, \varsigma) - \varsigma) d\delta F(\varsigma) + \int_m^x (g(\varsigma, m) - \varsigma) dF_1(\varsigma) = 0$$

$$(2.33)$$

where $g(x, \varsigma) \equiv F_2^{-1}(F_1(x) + \delta F(\varsigma))$.

LEMMA 2.1
$\mathbb{P}^* \in \mathcal{M}(\mathbb{P}^1, \mathbb{P}^2)$.

PROOF
(i) $S_1 \overset{\mathbb{P}^*}{\sim} \mathbb{P}^1$: direct.
(ii) $\mathbb{E}^{\mathbb{P}^*}[S_2|S_1] = S_1$: direct from the expression of q.
(iii) $S_2 \overset{\mathbb{P}^*}{\sim} \mathbb{P}^2$: Let $y > m$ be a point of the support of $\mathbb{P}^2$. Then $y = T_u(x)$ for some $x \geq m$, and the constraint $S_2 \overset{\mathbb{P}^*}{\sim} \mathbb{P}^2$ gives from (2.30)

$$dF_2(y) = q(x)dF_1(x) \qquad (2.34)$$

Let $y < m$ be a point of the support of $\mathbb{P}^2$. Then, $y = T_d(x)$ for some $x > m$, and the constraint $S_2 \overset{\mathbb{P}^*}{\sim} \mathbb{P}^2$ gives from (2.30)

$$d(\delta F)(y) = -(1 - q(x))dF_1(x) \qquad (2.35)$$

By subtracting equations (2.34, 2.35), we get

$$dF_2(T_u(x)) - d(\delta F)(T_d(x)) = dF_1(x) \qquad (2.36)$$

Equations (2.36, 2.35) are then equivalent to the system of ODEs (2.31)-(2.32) on $[m, \infty)$, with the boundary condition $T_u(m) = T_d(m) = m$. ☐

In Figure (2.4), we have plotted the maps T_d and T_u corresponding to two log-normal densities with variances 0.04 and 0.32 (increasing in convex order). Note that the expression (2.31) for the map $T_u(x)$ looks like the Fréchet–Hoeffding solution, except the presence of an additional term $\delta F(T_d(x))$ arising from our martingale condition. From Figure (2.4), we see that $\mathbb{P}^*$ is clearly left-monotone.

Primal side

We next introduce a remarkable triple of dual variables $(\lambda_1^*, \lambda_2^*, H^*)$ corresponding to a smooth coupling c. The dynamic hedging component H^* is

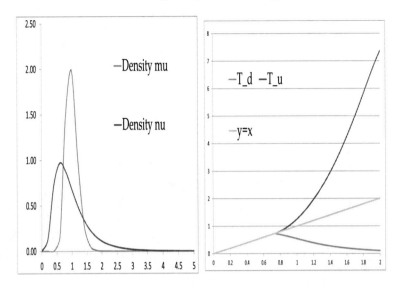

Figure 2.4: Maps T_d and T_u (right) built from two log-normal densities $\mathbb{P}^1 = \nu$ and $\mathbb{P}^2 = \mu$ (left) with variances 0.04 and 0.32. $m = 0.731$. $\mathbb{P}^*$ is left-monotone.

defined up to an arbitrary constant by:

$$H^{*'}(s_1) = \frac{c_{s_1}(s_1, T_u(s_1)) - c_{s_1}(s_1, T_d(s_1))}{T_u(s_1) - T_d(s_1)}, \quad \forall\, s_1 \geq m \tag{2.37}$$

The payoff function λ_2^* is defined up to an arbitrary constant by:

$$\begin{aligned}
\lambda_2^{*'}(s_2) &= c_{s_2}(T_u^{-1}(s_2), s_2) - H^* \circ T_u^{-1}(s_2), \quad \forall\, s_2 \geq m \tag{2.38}\\
&= c_{s_2}(T_d^{-1}(s_2), s_2) - H^* \circ T_d^{-1}(s_2), \quad \forall\, s_2 < m.
\end{aligned}$$

The corresponding function λ_1^* is given by: $\forall\, s_1 \in \mathbb{R}_+$,

$$\begin{aligned}
\lambda_1^*(s_1) &= \mathbb{E}^{\mathbb{P}^*}\big[c(S_1, S_2) - \lambda_2^*(S_2)|S_1 = s_1\big] \tag{2.39}\\
&= q(s_1)\big(c(s_1, .) - \lambda_2^*\big) \circ T_u(s_1) + \big(1 - q(s_1)\big)\big(c(s_1, .) - \lambda_2^*\big) \circ T_d(s_1)
\end{aligned}$$

THEOREM 2.8
*Assume that $\mathbb{P}^1 \leq \mathbb{P}^2$ are probability measures on $\mathbb{R}_+$ with first moments S_0 and assumption 4(**Assum**$(\mathbb{P}^1, \mathbb{P}^2)$) holds. Assume further that $\lambda_1^{*+} \in L^1(\mathbb{P}^1)$, $\lambda_2^{*+} \in L^1(\mathbb{P}^2)$, and that the partial derivative of the coupling $c_{s_1 s_2 s_2}$ exists and $c_{s_1 s_2 s_2} > 0$ on $\mathbb{R}_+^2$. Then:*
(i) $(\lambda_1^*, \lambda_2^*, H^*) \in \mathcal{M}^*(\mathbb{P}^1, \mathbb{P}^2)$,
(ii) the strong duality holds for the martingale transport problem, $\mathbb{P}^*$ is a

solution of $\widetilde{\mathrm{MK}}_2^*$, *and* $(\lambda_1^*, \lambda_2^*, H^*)$ *is a solution of* $\widetilde{\mathrm{MK}}_2$:

$$\mathbb{E}^{\mathbb{P}^*}[c(S_1, S_2)] = \widetilde{\mathrm{MK}}_2^* = \widetilde{\mathrm{MK}}_2 = \mathbb{E}^{\mathbb{P}^1}[\lambda_1^*(S_1)] + \mathbb{E}^{\mathbb{P}^2}[\lambda_2^*(S_2)].$$

The proof is briefly reported below and in Section 2.3 we expand our strategy to derive other optimal solutions.

PROOF see [95] for details From the weak duality, we have $\widetilde{\mathrm{MK}}_2^* \le \widetilde{\mathrm{MK}}_2$.

(i) As $\mathbb{P}^*$ defined by (2.30) belongs to $\mathcal{M}(\mathbb{P}^1, \mathbb{P}^2)$ (see Lemma 2.1), we have the inequality

$$\mathbb{E}^{\mathbb{P}^*}[c(S_1, S_2)] \le \widetilde{\mathrm{MK}}_2^*$$

(ii) If we assume that the primal is attained by a triplet $(\lambda_1^*, \lambda_2^*, H^*)$, we should have from (2.29) that for all $s \in \mathbb{R}_+$

$$\lambda_1^*(s) + \lambda_2^*(T_u(s)) + H^*(s)(T_u(s) - s) = c(s, T_u(s)) \qquad (2.40)$$
$$\lambda_1^*(s) + \lambda_2^*(T_d(s)) + H^*(s)(T_d(s) - s) = c(s, T_u(s)) \qquad (2.41)$$

Similarly, we should have that

$$\lambda_1^*(s_1) = \sup_{s_2 \in \mathbb{R}_+} \{c(s_1, s_2) - \lambda_2^*(s_2) - H^*(s_1)(s_2 - s_1)\}$$

From Equations (2.40, 2.41), the supremum above should be reached for $s_2 = T_d(s_1)$ and $s_2 = T_u(s_1)$. This gives two additional equations:

$$\partial_{s_2}\lambda_2^*(T_u(s)) + H^{*'}(s) = \partial_{s_2}c(s, T_u(s))$$
$$\partial_{s_2}\lambda_2^*(T_d(s)) + H^{*'}(s) = \partial_{s_2}c(s, T_d(s))$$

As for $s \ge m$, $T_u \in [m, \infty)$ and $T_d \in (0, m]$, the last two equations define λ_2^* up to an arbitrary constant. Equations (2.40-2.41) define then λ_1^* and H^*. Under the condition $c_{122} > 0$, one can show that $(\lambda_1^*, \lambda_2^*, H^*) \in \mathcal{M}^*(\mathbb{P}^1, \mathbb{P}^2)$. This implies that

$$\mathbb{E}^{\mathbb{P}^*}[c(S_1, S_2)] \le \widetilde{\mathrm{MK}}_2^* \le \widetilde{\mathrm{MK}}_2 \le \mathbb{E}^{\mathbb{P}^1}[\lambda_1^*(S_1)] + \mathbb{E}^{\mathbb{P}^2}[\lambda_2^*(S_2)]$$

(iii) We conclude the proof by checking with a straightforward computation that

$$\mathbb{E}^{\mathbb{P}^1}[\lambda_1^*(S_1)] + \mathbb{E}^{\mathbb{P}^2}[\lambda_2^*(S_2)] = \mathbb{E}^{\mathbb{P}^*}[c(S_1, S_2)]$$

□

Table 2.2: OT versus MOT. $\lambda^*(s_1) \equiv \sup_{s_2}\{c(s_1, s_2) - \lambda(s_2)\}$ and $\lambda^{**}(\cdot)$ denotes the concave envelope of $\lambda(\cdot)$ which is the smallest concave function greater than or equal to u.

OT, MK$_2$	MOT, $\widetilde{\text{MK}}_2$
$\sup_{\mathbb{P}\,:\,S_1\sim\mathbb{P}^1,S_2\sim\mathbb{P}^2}\mathbb{E}^{\mathbb{P}}[c(S_1, S_2)]$	$\sup_{\mathbb{P}\,:\,S_1\sim\mathbb{P}^1,S_2\sim\mathbb{P}^2,\mathbb{E}^{\mathbb{P}}[S_2|S_1]=S_1}\mathbb{E}^{\mathbb{P}}[c(S_1, S_2)]$
$\inf_{\lambda_1,\lambda_2}\mathbb{E}^{\mathbb{P}^1}[\lambda_1(S_1)] + \mathbb{E}^{\mathbb{P}^2}[\lambda_2(S_2)]$	$\inf_{\lambda_1,\lambda_2,H}\mathbb{E}^{\mathbb{P}^1}[\lambda_1(S_1)] + \mathbb{E}^{\mathbb{P}^2}[\lambda_2(S_2)]$
$\lambda_1(s_1) + \lambda_2(s_2) \geq c(s_1, s_2)$	$\lambda_1(s_1) + \lambda_2(s_2) + H(s_1)(s_2 - s_1) \geq c(s_1, s_2)$
$\inf_{\lambda}\mathbb{E}^{\mathbb{P}^1}[\lambda^*(S_1)] + \mathbb{E}^{\mathbb{P}^2}[\lambda(S_2)]$	$\inf_{\lambda}\mathbb{E}^{\mathbb{P}^1}[(c(S_1, \cdot) - \lambda(\cdot))^{**}(S_1)] + \mathbb{E}^{\mathbb{P}^2}[\lambda(S_2)]$
(co)-monotone map: $c_{12} > 0$	(co)-monotone martingale map: $c_{122} > 0$
Primal/dual attained	Dual attained

REMARK 2.6 In comparison with the Fréchet–Hoeffding solution, our martingale solution is not symmetric in s_1 and s_2. This is due to the martingale condition (and the convex order assumption on $\mathbb{P}^1$ and $\mathbb{P}^2$) that breaks this invariance. ⬜

Example 2.8 $c(s_1, s_2) = \ln^2 \frac{s_2}{s_1}$
The upper bound is

$$\widetilde{\text{MK}}_2 = \int_0^\infty \frac{(T_u(x) - x)\ln^2 \frac{T_d(x)}{x} + (x - T_d(x))\ln^2 \frac{T_u(x)}{x}}{T_u(x) - T_d(x)}\mathbb{P}^1(dx)$$

⬜

2.2.4   OT versus MOT: A summary

We summarize our presentation by comparing various results in OT and in the martingale counterpart (see Table (2.2)). We have seen that OT is linked to the Hamilton–Jacobi equation. In MOT (in particular in our approach to the Skorokhod embedding problem), Hamilton–Jacobi–Bellman PDEs appear naturally.

2.2.5   Martingale Brenier's solution

The enormous development of OT in the last decades was initiated by Brenier's celebrated theorem, briefly reviewed in Theorem 2.3. Hence a most natural question is to obtain similar results also for the martingale version of the transport problem. The literature on this topic includes [82, 83]. This seems a potentially very interesting problem for mathematicians working in OT to tackle this problem, particularly in $\mathbb{R}^d$.

We briefly state below MOT in $\mathbb{R}^d_+$. We denote $\mathbb{P}^1$ and $\mathbb{P}^2$ the marginals of S_1 and S_2 in $\mathbb{R}^d_+$ and S^i_1 the i-component of S_1. The knowledge of marginals $\mathbb{P}^1$ and $\mathbb{P}^2$ is not very common in finance as the (known) marginals are usually one-dimensional (e.g. Vanillas), see however our discussion in Section 2.1.3. A notable exception arises in fixed income and foreign exchange markets (see Example 2.1) where Vanillas on spread swap rates, i.e., $(S_2 - KS_1)^+$, are quoted on the market.
MOT reads

$$\widetilde{\mathrm{MK}}_2 = \inf_{\lambda_1 \in L^1(\mathbb{P}^1), \lambda_2 \in L^1(\mathbb{P}^2), (H^i(\cdot))_{1 \leq i \leq d}} \mathbb{E}^{\mathbb{P}^1}[\lambda_1(S_1)] + \mathbb{E}^{\mathbb{P}^2}[\lambda_2(S_2)]$$

such that $\lambda_1(s_1) + \lambda_2(s_2) + \sum_{i=1}^d H^i(s_1)(s^i_2 - s^i_1) \geq c(s_1, s_2), \quad \forall\, (s_1, s_2) \in (\mathbb{R}^d_+)^2$. Taking for granted that the primal is attained (the dual is attained by weak compactness), the (strong) duality result implies as before that

$$\lambda_1(s_1) + \lambda_2(s_2) + \sum_{i=1}^d H^i(s_1)(s^i_2 - s^i_1) = c(s_1, s_2), \quad \mathbb{P}^* - \text{a.s.}$$

We have $d + 2$ unknown functions $(\lambda_1, \lambda_2, (H^i(\cdot))_{1 \leq i \leq d})$ (defined on (a subset of) $\mathbb{R}^d_+$) and it is tempting to guess that the optimal martingale measure $\mathbb{P}^*$ is localized on some maps $(T^\alpha)_{\alpha=1,\ldots,N}$. For each map - denoted schematically by T with components $(T_1, \ldots, T_d)$ - we should have: $\forall\, s_1 \in \mathbb{R}^d$,

$$\lambda_1(s_1) + \lambda_2(T(s_1)) + \sum_{i=1}^d H^i(s_1)(T_i(s_1) - s^i_1) = c(s_1, T(s_1)) \quad (2.42)$$

$$\partial_{s^i_2} \lambda_2(T(s_1)) + H^i(s_1) = \partial_{s^i_2} c(s_1, T(s_1)), \quad \forall\, i = 1, \ldots, d, \quad (2.43)$$

On the dual side, we should have :

$$\mathbb{P}^*(ds_1, s_2) = \sum_{\alpha=1}^N q_\alpha(s_1) \delta_{T^\alpha(s_1)}(ds_2) \mathbb{P}^1(ds_1)$$

where the functions $(q_\alpha)_{\alpha=1,\ldots,N}$ are constrained by the algebraic equations:

$$\sum_{\alpha=1}^N q_\alpha(s_1) = 1, \quad \sum_{\alpha=1}^N q_\alpha(s_1)(T^\alpha(s_1) - s_1) = 0 \quad (2.44)$$

An open problem is to extend our approach (or using a different viewpoint, e.g. cyclical monotonicity) in $d = 1$ to $\mathbb{R}^d_+$. Notable progress has been made in [82, 57, 120].
The maps $(T^\alpha)_{\alpha=1,\ldots,N}$ could not be surjective in $\mathbb{R}^d_+$, indicating that the maps span $\mathbb{R}^d_+$ with a complicated geometry. Indeed, if it was the case, we would have $2N + 1 + d$ equations (2.42, 2.43, 2.44) for $dN + N + 2 + d$ unknown scalar functions $f : \mathbb{R}^d_+ \mapsto \mathbb{R}$ (i.e., $(T^\alpha), (q_\alpha), \lambda_1, \lambda_2, (H^i)$). Requiring the number of equations to be greater than or equal to the number of unknowns imply that $N(1 - d) \geq 1$. This is only possible in $d = 1$.

2.2.6　Symmetries in MOT

2.2.6.1　Martingale Spence–Mirrlees condition

The martingale counterpart of the Spence–Mirrlees condition is $c_{s_1 s_2 s_2} > 0$. This condition is natural in the present setting. Indeed, the optimization problem should not be affected by the modification of the coupling from c to $\bar{c}(s_1, s_2) \equiv c(s_1, s_2) + \Lambda_1(s_1) + \Lambda_1(s_2) + H(s_1)(s_2 - s_1)$ for any $\Lambda_1 \in L^1(\mathbb{P}^1)$, $\Lambda_2 \in L^1(\mathbb{P}^2)$, and $H \in L^0$, the new optimal cost being $\widetilde{\mathrm{MK}}_2 + \mathbb{E}^{\mathbb{P}^1}[\Lambda_1(S_1)] + \mathbb{E}^{\mathbb{P}^2}[\Lambda_2(S_2)]$. Since $\bar{c}_{s_1 s_2 s_2} = c_{s_1 s_2 s_2}$, it follows that the condition $c_{s_1 s_2 s_2} > 0$ is stable for the above transformation of the coupling. In particular, note that the original Spence–Mirrlees condition is not preserved under this transformation: $\bar{c}_{s_1 s_2} = c_{s_1 s_2} + H'(s_1) \neq c_{s_1 s_2}$.

2.2.6.2　Mirror coupling: The right-monotone martingale transport plan

Suppose that $c_{s_1 s_2 s_2} < 0$. Then, the upper bound $\widetilde{\mathrm{MK}}_2$ is attained by the right-monotone martingale transport map

$$\mathbb{P}_*(ds_1, ds_2) = \mathbb{P}^1(ds_1) \left(q(s_1)\delta_{\bar{T}_u(s_1)}(ds_2) + (1 - q(s_1))\delta_{\bar{T}_d(s_1)}(ds_2) \right),$$

$$q(x) = \frac{x - \bar{T}_d(x)}{\bar{T}_u(x) - \bar{T}_d(x)}$$

where $(\bar{T}_d, \bar{T}_u)$ is defined as in (2.31, 2.32) with the pair of probability measures $(\bar{\mathbb{P}}^1, \bar{\mathbb{P}}^2)$:

$$\bar{F}^1(s_1) \equiv 1 - F^1(-s_1), \text{ and } \bar{F}^2(s_2) \equiv 1 - F^2(-s_2).$$

To see this, we rewrite the OT problem equivalently with modified inputs:

$$\bar{c}(s_1, s_2) \equiv c(-s_1, -s_2), \quad \bar{\mathbb{P}}^1\big((-\infty, s_1]\big) \equiv \mathbb{P}^1\big([-s_1, \infty)\big),$$
$$\bar{\mathbb{P}}^2\big((-\infty, s_2]\big) \equiv \mathbb{P}^2\big([-s_2, \infty)\big),$$

so that $\bar{c}_{s_1 s_2 s_2} > 0$, as required in Theorem 2.8. Note that the martingale constraint is preserved by the map $(s_1, s_2) \mapsto (-s_1, -s_2)$ (and not by our parity transformation $(s_1, s_2) \mapsto (s_1, -s_2)$ in OT).

Suppose that $c_{s_1 s_2 s_2} > 0$. Then, the lower bound problem is explicitly solved by the right-monotone martingale transport plan. Indeed, it follows from the first part of the present remark that:

$$\inf_{\mathbb{P} \in \mathcal{M}(\mathbb{P}^1, \mathbb{P}^2)} \mathbb{E}^{\mathbb{P}}\big[c(S_1, S_2)\big] = - \sup_{\mathbb{P} \in \mathcal{M}(\mathbb{P}^1, \mathbb{P}^2)} \mathbb{E}^{\mathbb{P}}\big[-c(S_1, S_2)\big]$$

$$= - \sup_{\mathbb{P} \in \mathcal{M}(\mathbb{P}^1, \mathbb{P}^2)} \mathbb{E}^{\mathbb{P}}\big[-\bar{c}(-S_1, -S_2)\big]$$

$$= - \sup_{\mathbb{P} \in \mathcal{M}(\bar{\mathbb{P}}^1, \bar{\mathbb{P}}^2)} \mathbb{E}^{\mathbb{P}}\big[-\bar{c}(S_1, S_2)\big]$$

$$= \mathbb{E}^{\mathbb{P}_*}\big[c(S_1, S_2)\big]$$

This parity transformation exchanges the left and right-monotone martingale transport plan where the marginals have support in $\mathbb{R}$.

2.2.6.3 Change of numéraire

We define the involution $\mathcal{S}$ [34] (i.e., $\mathcal{S}^2 = \mathrm{Id}$) on a payoff function c by

$$(\mathcal{S}c)(s_1, s_2) \equiv s_2 c(\frac{1}{s_1}, \frac{1}{s_2})$$

We have

$$\sup_{\mathbb{P}\in\mathcal{M}(\mathbb{P}^1,\mathbb{P}^2)} \mathbb{E}^{\mathbb{P}}[(\mathcal{S}c)(S_1, S_2)] = \sup_{\mathbb{P}\in\mathcal{M}(\mathbb{P}^1,\mathbb{P}^2)} \mathbb{E}^{\mathbb{P}}[S_2 c\left(\frac{1}{S_1}, \frac{1}{S_2}\right)]$$

$$= S_0 \sup_{\mathbb{Q}\in\mathcal{M}(\mathcal{S}(\mathbb{P}^1),\mathcal{S}(\mathbb{P}^2))} \mathbb{E}^{\mathbb{Q}}[c(\bar{S}_{t_1}, \bar{S}_{t_2})]$$

where $\mathcal{S}(\mathbb{P}^i)$, $i = 1, 2$ has a density $(\mathcal{S}f^i)(s) = \frac{1}{S_0 s^3} f^i(\frac{1}{s})$ where f^i the density of $\mathbb{P}^i$. We have used that by working in the numéraire associated to the discrete martingale S_t:

$$\mathbb{E}^{\mathbb{P}}[S_2 c\left(\frac{1}{S_1}, \frac{1}{S_2}\right)] = S_0 \mathbb{E}^{\mathbb{Q}}[c\left(\frac{1}{S_1}, \frac{1}{S_2}\right)]$$

with $\frac{d\mathbb{Q}}{d\mathbb{P}}|_{\mathcal{F}_{t_i}} = \frac{S_i}{S_0}$. Under $\mathbb{Q}$, $\frac{1}{S_i}$ is a discrete martingale: $\mathbb{E}^{\mathbb{Q}}[\frac{1}{S_2}|\frac{1}{S_1}] = \frac{1}{S_1}$. This involution $\mathcal{S}$ satisfies

$$(\mathcal{S}c)_{122}(s_1, s_2) = -\frac{1}{s_1^2 s_2^3} c_{122}(\frac{1}{s_1}, \frac{1}{s_2})$$

and exchanges therefore the left and right-monotone martingale transport plan where the marginals have support in $\mathbb{R}_+$.

2.2.7 c-cyclical monotonicity

An important concept in OT is the notion of c-cyclical monotonicity: roughly speaking, a transport plan is optimal for a given transport problem if all "cycles" in its support set satisfy a particular monotonicity property. This result is often fundamental to characterizing optimal solutions, see [85]. There is a significant line of research about versions and applications of this idea in the martingale transport context ([18, 16, 134, 15, 144, 86, 83]). This approach is somewhat parallel to our ODE approach presented in the previous sections.

2.2.8 Martingale McCann's interpolation

In this section, we recall McCann's interpolation theory and present a martingale extension. Our reminder follows closely Chapter 5 in Villani's book [139].

Here we consider the case $\mathbb{R}^d_+$, although our martingale extension focuses only on the real line. Two probability measures $\mathbb{P}^0$ and $\mathbb{P}^1$ (not necessarily in convex order here) can be (trivially) linearly interpolated by $\mathbb{P}_t = (1-t)\mathbb{P}^0 + t\mathbb{P}^1$ with $t \in [0,1]$.

In mathematical physics, one often considers minimisation of functionals on the space of probability measures $\mathcal{P}(\mathbb{R}^d_+)$ - see similar problems in mathematical finance in Section 2.1.6: $\inf_{\mathbb{P} \in \mathcal{P}(\mathbb{R}^d_+)} \mathcal{F}(\mathbb{P})$. The proof of the existence of a unique minimiser is greatly simplified if we can prove that the functional $\mathcal{F}$ is strictly convex. However classical examples such as

$$\mathcal{F}_2(\mathbb{P}) \equiv \int_{\mathbb{R}^d_+ \times \mathbb{R}^d_+} W(x-y)\mathbb{P}(dx)\mathbb{P}(dy) \qquad (2.45)$$

do not satisfy this property even if W is strictly convex. Indeed, a straightforward computation gives

$$\frac{d^2}{dt^2}\mathcal{F}_2(\mathbb{P}_t) = 2\int W(x-y)\left(\mathbb{P}^0(dx) - \mathbb{P}^1(dx)\right)\left(\mathbb{P}^0(dy) - \mathbb{P}^1(dy)\right)$$

for which it is not possible to conclude the convex property. This was McCann's original motivation [114] for introducing a new notion of probability interpolation and convexity which can handle the above example. His approach is strongly linked to Brenier's theorem in OT as described below.

Let's take a strictly concave cost $c = c(s_1 - s_2)$ on $\mathbb{R}^d_+$. According to Brenier's theorem, there exists a function u such that (see Equation (2.13))

$$[\mathrm{Id} - \nabla(c^*(\nabla u))]\#\mathbb{P}^0 = \mathbb{P}^1$$

where $T\#\mathbb{P}^0 = \mathbb{P}^1$ denotes the forward image of $\mathbb{P}^0$ by the map T. For T differentiable, this is equivalent to $\mathbb{P}^0(x) = |\det \nabla T|\mathbb{P}^1(T(x))$.

McCann's displacement interpolation $[\mathbb{P}^0, \mathbb{P}^1]_t$ of two probability measures $\mathbb{P}^0$ and $\mathbb{P}^1$ is then defined by

$$[\mathbb{P}^0, \mathbb{P}^1]_t \equiv [\mathrm{Id} - t\nabla(c^*(\nabla u))]\#\mathbb{P}^0, \quad t \in [0,1]$$

Note that by construction $[\mathbb{P}^0, \mathbb{P}^1]_{t=0} = \mathbb{P}^0$ and $[\mathbb{P}^0, \mathbb{P}^1]_{t=1} = \mathbb{P}^1$.

DEFINITION 2.8 Convex displacement
(i) *A subset $\mathcal{P}$ of $\mathcal{P}(\mathbb{R}^d_+)$ is said to be displacement convex if it is stable under displacement interpolation:* $\mathbb{P}^0, \mathbb{P}^1 \in \mathcal{P} \Longrightarrow [\mathbb{P}^0, \mathbb{P}^1]_t \in \mathcal{P}, \quad \forall\, t \in [0,1]$.
(ii) *A functional $\mathcal{F}$ defined on a displacement convex subset $\mathcal{P}$ is displacement (resp. strictly) convex if for all $\mathbb{P}^0, \mathbb{P}^1$ in $\mathcal{P}$, the function $t \mapsto \mathcal{F}([\mathbb{P}^0, \mathbb{P}^1]_t)$ is (resp. strictly) convex on $[0,1]$.*

In particular, the functional $\mathcal{F}_2$ is displacement convex if W is convex [114]. One can then show the existence of a unique minimiser of $\mathcal{F}$ using this displacement convex property (see [114]).

For applications in mathematical finance, one can cite the problem of minimisations of functionals defined on a subset of $\mathcal{P}(\mathbb{R}_+)$ totally ordered with respect to the convex order (see Definition 2.6). Note that in comparison with the usual convex interpolation, McCann's interpolation does not preserve the convex order property: If $\mathbb{P}^0 \leq \mathbb{P}^1$, then we don't have $\mathbb{P}^0 \leq [\mathbb{P}^0, \mathbb{P}^1]_t \leq \mathbb{P}^1$. It is therefore not applicable in our present context.

Using the framework of MOT, we define the convex order interpolation of two measures $\mathbb{P}^0 \leq \mathbb{P}^1$ defined on $\mathbb{R}_+$ as

DEFINITION 2.9 Martingale convex interpolation

$$[\mathbb{P}^0, \mathbb{P}^1]_t = \text{Law}(S_t), \quad t \in [0, 1]$$

where the random variable S_t is defined by a two step Markov chain: $S_t \equiv S(1 - t) + tT_u(S)$ with probability $q(s) \equiv \frac{s - T_d(s)}{T_u(s) - T_d(s)}$, $S_t \equiv S(1 - t) + tT_d(S)$ with probability $1 - q(s)$ and $S \sim \mathbb{P}^0$.

We still use the same bracket notation $[\mathbb{P}^0, \mathbb{P}^1]_t$. There should not be any confusion as we are focusing now only on this martingale convex interpolation. We have

LEMMA 2.2

$$\mathbb{P}^0 = [\mathbb{P}^0, \mathbb{P}^1]_{t=0} \leq [\mathbb{P}^0, \mathbb{P}^1]_t \leq [\mathbb{P}^0, \mathbb{P}^1]_{t=1} = \mathbb{P}_1$$

for all $t \in [0, 1]$.

PROOF We set $C(t, K) \equiv \mathbb{E}^{[\mathbb{P}^0, \mathbb{P}^1]_t}[(S_t - K)^+]$. We have that

$$C(t, K) = \mathbb{E}^{\mathbb{P}^0}[\ q(S)(S(1 - t) + tT_u(S) - K)^+ \\ + (1 - q(S))(S(1 - t) + tT_d(S) - K)^+]$$

By differentiating with respect to t, we get

$$\partial_t \mathbb{E}[(S_t - K)^+] = \mathbb{E}^{\mathbb{P}^0}[\frac{(T_u(S) - S)(S - T_d(S))}{T_u(S) - T_d(S)} \\ \left(1_{S(1-t)+tT_u(S)>K} - 1_{S(1-t)+tT_d(S)>K}\right)]$$

which is obviously positive. $\qquad\square$

Note that the linear interpolation $\mathbb{P}_t = (1 - t)\mathbb{P}^0 + t\mathbb{P}^1$ preserves also the convex order property. We explain in the next section why our interpolation seems better.

2.2.8.1   Application

We introduce a subset $\mathcal{P}_{c.o.}$ of the space of probability measures, absolutely continuous with respect to the Lebesgue measure on $\mathbb{R}_+$, totally ordered with respect to the convex order property. This space is stable under convex order displacement interpolation. Note that $\mathcal{P}_{c.o.}$ is a convex set in the usual sense.

DEFINITION 2.10 Martingale convex displacement   *We say that a functional $\mathcal{F}$ on $\mathcal{P}_{c.o.}$ is displacement (resp. strictly) convex order if the map $t \mapsto \mathcal{F}([\mathbb{P}^0, \mathbb{P}^1]_t)$ is (resp. strictly) convex on $[0,1]$.*

The functional $\mathcal{F}_2$ (2.45) introduced above is an example of displacement (resp. strictly) convex order functional as proved in Theorem 2.9. This example is not convex in the usual sense (i.e., for $\mathbb{P}_t = (1-t)\mathbb{P}^0 + t\mathbb{P}^1$). This is why our notion seems better. Our definition (in particular the fact that q remains time independent under convex order interpolation) seems to be the right definition. As an application of the convex order interpolation, we have

THEOREM 2.9 see a similar theorem in OT, Theorem 5.32 [139]
Consider the functional $\mathcal{F}_2(\mathbb{P})$ defined on $\mathcal{P}_{c.o.}$ (see the definition above) where W is strictly convex. Then, there is at most one minimiser for $\mathcal{F}_2(\mathbb{P})$ on $\mathcal{P}_{c.o.}$.

PROOF   (1) $\mathcal{F}_2(\mathbb{P})$ is strictly convex with respect to the martingale convex interpolation. Indeed, set X, Y two independent r.v. with law $\mathbb{P}^0$. Then,

$$\mathcal{F}_2(\mathbb{P}_t) = \mathbb{E}[W(X(1-t) + tT_u(X) - Y(1-t) - tT_u(Y))q(X)q(Y)]$$
$$+ \mathbb{E}[W(X(1-t) + tT_d(X) - Y(1-t) - tT_d(Y))(1-q(X))(1-q(Y))]$$
$$+ \mathbb{E}[W(X(1-t) + tT_d(X) - Y(1-t) - tT_u(Y))(1-q(X))q(Y)]$$
$$+ \mathbb{E}[W(X(1-t) + tT_u(X) - Y(1-t) - tT_d(Y))q(X)(1-q(Y))]$$

By differentiating twice with respect to t, we get our result if W is strictly convex.
(2) Let $\mathbb{P}^0$ and $\mathbb{P}^1$ be two distinct minimizers, and consider $[\mathbb{P}^0, \mathbb{P}^1]_{\frac{1}{2}}$. The strict convexity of $\mathcal{F}_2(\mathbb{P})$ implies that

$$\mathcal{F}_2([\mathbb{P}^0, \mathbb{P}^1]_{\frac{1}{2}}) < \frac{1}{2}\left(\mathcal{F}_2(\mathbb{P}^0) + \mathcal{F}_2(\mathbb{P}^1)\right)$$

which is impossible since $\mathbb{P}^0$ and $\mathbb{P}^1$ are minimizers.   ⬜

2.2.9   Multi-marginals extension

A natural extension of our 2-period MOT is to consider the multi-marginal case $(n > 2)$:

DEFINITION 2.11

$$\widetilde{MK}_n \equiv \inf_{\mathcal{M}^*(\mathbb{P}^1,\dots \mathbb{P}^n)} \sum_{i=1}^n \mathbb{E}^{\mathbb{P}^i}[\lambda_i(S_i)] \qquad (2.46)$$

where $\mathcal{M}^*(\mathbb{P}^1,\dots,\mathbb{P}^n)$ is the set of functions $\lambda_i \in L^1(\mathbb{P}^i)$ and $(H_i)_{1 \leq i \leq n-1}$ bounded continuous functions on $\mathbb{R}_+^i$ such that

$$\sum_{i=1}^n \lambda_i(s_i) + \sum_{i=1}^{n-1} H_i(s_1,\dots,s_i)(s_{i+1} - s_i) \geq c(s_1,\dots,s_n) \qquad (2.47)$$

for all $(s_1,\dots,s_n) \in \mathbb{R}_+^n$.

The dual formulation is (the proof is the same as in Theorem 2.7)

THEOREM 2.10

$$\widetilde{MK}_n = \sup_{\mathbb{P} \in \mathcal{M}(\mathbb{P}^1,\dots,\mathbb{P}^n)} \mathbb{E}^{\mathbb{P}}[c(S_1,\dots,S_n)]$$

where

$$\mathcal{M}(\mathbb{P}^1,\dots,\mathbb{P}^n) \equiv \{\mathbb{P} \; : \; S_i \overset{\mathbb{P}}{\sim} \mathbb{P}^i, \mathbb{E}^{\mathbb{P}}_{t_{i-1}}[S_i] = S_{i-1}, i = 1,\dots,n\} \qquad (2.48)$$

Here the operator $\mathbb{E}^{\mathbb{P}}_{t_{i-1}}[\cdot] \equiv \mathbb{E}^{\mathbb{P}}[\cdot|S_{i-1},\dots,S_1,S_0]$.

REMARK 2.7 Markov MOT   If we assume that $H_i = H_i(s_i)$ in (2.47) then in the dual formulation, $\mathcal{M}(\mathbb{P}^1,\dots,\mathbb{P}^n)$ is replaced by

$$\mathcal{M}^{\text{Markov}}(\mathbb{P}^1,\dots,\mathbb{P}^n) \equiv \{\mathbb{P} \; : \; S_i \overset{\mathbb{P}}{\sim} \mathbb{P}^i, \mathbb{E}^{\mathbb{P}}[S_i|S_{i-1}] = S_{i-1}, i = 1,\dots,n\}$$

, i.e., $\mathbb{P} \in \mathcal{M}^{\text{Markov}}$ satisfies the Markov property.   ⬜

This problem was solved in the classical OT problem in [39, 78] (see Section 2.1.9). In our present martingale version, by using a Markov property, and a specific class of cost functions, $c = \sum_{i=1}^{n-1} c^i(s_i, s_{i+1})$ with $c^i_{s_i s_{i+1} s_{i+1}} > 0$, it is natural to guess that the optimal martingale measure will be supported by $2 \times n$ maps (T_d^i, T_u^i) satisfying ODEs (2.31,2.32) with F_1 and F_2 replaced by F_i and F_{i+1}. More precisely, we have

THEOREM 2.11 see [95]
*Suppose $\mathbb{P}^1 \leq \dots \leq \mathbb{P}^n$ in convex order, with finite first moment S_0. $\mathbb{P}^1,\dots,\mathbb{P}^{n-1}$ have no atoms, and let Assumption 4 (**Assum**($\mathbb{P}^i, \mathbb{P}^{i+1}$)) holds true for $\delta F = F_{i+1} - F_i$, for all $1 \leq i \leq n-1$. Assume further that*

(i) c^i *have linear growth, that the cross derivatives* c^i_{xyy} *exist and satisfy* $c^i_{xyy} > 0$,

(ii) $\lambda_i^{1*}, \lambda_i^{2*}$ *satisfy the integrability conditions* $(\lambda_i^{1*})^+ \in L^1(\mathbb{P}^i)$, $(\lambda_i^{2*})^+ \in L^1(\mathbb{P}^{i+1})$. $H_i^*, \lambda_i^{1*}, \lambda_i^{2*}$ *are defined as in (2.37, 2.38, 2.39) with* c^i *substituted to* c *and* (T_u^i, T_d^i) *substituted to* (T_u, T_d).

Then, the strong duality holds, the transference map

$$\mathbb{P}^*(ds_1, \ldots, ds_n) = \mathbb{P}^1(ds_1) \prod_{i=1}^{n-1} \Big(q_i(s_i) \delta_{T_u^i(s_i)}(ds_{i+1}) + (1 - q_i(s_i)) \delta_{T_d^i(s_i)}(ds_{i+1}) \Big)$$

is optimal for the martingale transport problem $\widetilde{\mathrm{MK}}_n$, *and* (λ^*, H^*) *is optimal for the primal problem, i.e.,*

$$\mathbb{E}^{\mathbb{P}^*}[c(S_1, \ldots, S_n)] = \sum_{i=1}^{n} \mathbb{E}^{\mathbb{P}^i}[\lambda_i^*].$$

with $\lambda_i^*(s) \equiv 1_{i<n} \lambda_i^{1*}(s) + 1_{i>1} \lambda_{i-1}^{2*}(s)$, $\quad i = 1, \ldots, n$.

The proof is similar to the two periods case and therefore it is not reported.

Example 2.9 Discrete-monitored variance swap

As a useful application in mathematical finance, this result gives the robust lower (see mirror coupling, Section 2.2.6.2) and upper bounds of a discrete-monitored variance swap delivering the discrete realized variance of log-returns at a maturity T:

$$-\frac{2}{T} \sum_{i=0}^{n-1} \ln^2 \left(\frac{s_{i+1}}{s_i} \right)$$

See Example 2.8 for $n = 2$. We have compared these bounds against market values denoted VS_{mkt} for the DAX index (2-Feb-2013) and different maturities (see Table 2.3). We report the prices $(\widetilde{\mathrm{MK}}_n)^{\frac{1}{2}} \times 100$. Note that for maturities less than 1.4 years, our upper bound is below the market price, highlighting an arbitrage opportunity. In practice, this arbitrage disappears if we include transaction costs for trading Vanilla options with low/high strikes. Moreover, we have assumed that Vanilla options with all maturities are traded, which is an idealization. As only liquid maturities - say $(T_i)_{i=1,\ldots,N}$ - are quoted, the Vanillas with illiquid maturities are obtained by linear interpolation of the variance: For all $t \in (T_i, T_{i+1})$, the implied volatility $\sigma_{BS}(t, K)$ for strike K and maturity t is set to

$$\sigma_{BS}(t, K)^2 t = \left(\sigma_{BS}(T_2, K)^2 T_2 - \sigma_{BS}(T_1, K)^2 T_1 \right) \frac{(t - T_1)}{T_2 - T_1} + \sigma_{BS}(T_1, K)^2 T_1$$

□

Table 2.3: $(\widetilde{\mathrm{MK}}_n)^{\frac{1}{2}} \times 100$ for discrete-monitored variance swap as a function of the maturity. Lower/upper bounds versus market prices $\mathrm{VS_{mkt}}$ (quoted in volatility $\times 100$).

Maturity (years)	$\mathrm{VS_{mkt}}$	Upper	Lower
0.4	18.47	18.45	16.73
0.6	19.14	18.70	17.23
0.9	20.03	19.63	17.89
1.4	21.77	21.62	19.03
1.9	22.89	23.06	19.63

2.2.10 Robust quantile hedging

We set $p \in [0, 1]$ and define the robust martingale quantile hedging price as

DEFINITION 2.12

$$\widetilde{\mathrm{MK}}_2(p) \equiv \inf_{\mathcal{M}_p^*(\mathbb{P}^1, \mathbb{P}^2)} \mathbb{E}^{\mathbb{P}^1}[\lambda_1(S_1)] + \mathbb{E}^{\mathbb{P}^2}[\lambda_2(S_2)] \qquad (2.49)$$

where $\mathcal{M}_p^*(\mathbb{P}^1, \mathbb{P}^2)$ is the set of continuous functions $\lambda_1 \in \mathrm{L}^1(\mathbb{P}^1)$, $\lambda_2 \in \mathrm{L}^1(\mathbb{P}^2)$ and H a bounded continuous function on $\mathbb{R}_+$ such that

$$\inf_{\mathbb{P} \in \mathcal{P}(\mathbb{R}_+^2)} \mathbb{P}[\lambda_1(S_1) + \lambda_2(S_2) + H(S_1)(S_2 - S_1) \geq c(S_1, S_2)] \geq p \qquad (2.50)$$

and $\lambda_1(s_1) + \lambda_2(s_2) + H(s_1)(s_2 - s_1) \geq 0$, $\forall (s_1, s_2) \in \mathbb{R}_+^2$.

Here $\mathcal{P}(\mathbb{R}_+^2)$ denotes the set of probability measures on $\mathbb{R}_+^2$ (see Remark 2.3 for an explanation of our choice $\mathbb{P} \in \mathcal{P}(\mathbb{R}_+^2)$ and not $\mathbb{P} \in \mathcal{M}(\mathbb{P}^1, \mathbb{P}^2)$ or $\mathbb{P} \in \mathcal{P}(\mathbb{P}^1, \mathbb{P}^2)$). We could have replaced the constraint on the trader's portfolio value by

$$\lambda_1(s_1) + \lambda_2(s_2) + H(s_1)(s_2 - s_1) \geq -L$$

where L is a threshold.

THEOREM 2.12 see [12]
Assume that $\mathbb{P}^1 \leq \mathbb{P}^2$ are probability measures on $\mathbb{R}_+$ with first moments S_0 and Assumption 3 holds. Define

$$\mathcal{A}(p) \equiv \{A \in \mathcal{F} \text{ closed} : \inf_{\mathbb{P} \in \mathcal{P}(\mathbb{R}_+^2)} \mathbb{P}[A] \geq p\}$$

Then the following holds:

$$\widetilde{\mathrm{MK}}_2^*(p) \equiv \inf_{A \in \mathcal{A}(p)} \sup_{\mathbb{P} \in \mathcal{M}(\mathbb{P}^1, \mathbb{P}^2)} \mathbb{E}^{\mathbb{P}}[c(S_1, S_2)1_A]$$

$$= \widetilde{\mathrm{MK}}_2(p)$$

PROOF

(i): For $A \in \mathcal{A}(p)$, we set $\widetilde{\mathrm{MK}}_2(A) = \sup_{\mathbb{P} \in \mathcal{M}(\mathbb{P}^1, \mathbb{P}^2)} \mathbb{E}^{\mathbb{P}}[c(S_1, S_2)1_A]$. From Theorem 2.1,

$$\widetilde{\mathrm{MK}}_2(A) = \inf\{\sum_{i=1}^2 \mathbb{E}^{\mathbb{P}^i}[\lambda_i(S_i)] \; : \; \sum_{i=1}^2 \lambda_i(s_i) + H(s_1)(s_2 - s_1) \geq c(s_1, s_2)1_A\}$$

$$\geq \widetilde{\mathrm{MK}}_2(p)$$

By taking the infimum over $A \in \mathcal{A}(p)$, we get

$$\widetilde{\mathrm{MK}}_2^*(p) \geq \widetilde{\mathrm{MK}}_2(p)$$

(ii): For $\epsilon > 0$, let $\sum_{i=1}^2 \mathbb{E}^{\mathbb{P}^i}[\lambda_i(S_i)] \in [\widetilde{\mathrm{MK}}_2(p), \widetilde{\mathrm{MK}}_2(p) + \epsilon)$ be such that there exists H bounded continuous satisfying

$$\inf_{\mathbb{P} \in \mathcal{P}(\mathbb{R}_+^2)} \mathbb{P}[\pi \equiv \sum_{i=1}^2 \lambda_i(S_i) + H(S_1)(S_2 - S_1) \geq c(S_1, S_2)] \geq p \qquad (2.51)$$

Define $A \equiv \{\pi \geq c(S_1, S_2)\}$. Then $A \in \mathcal{A}(p)$ and $\pi \geq c1_A$. Hence,

$$\widetilde{\mathrm{MK}}_2(p) + \epsilon \geq \sum_{i=1}^2 \mathbb{E}^{\mathbb{P}^i}[\lambda_i(S_i)] = \sup_{\mathbb{P} \in \mathcal{M}(\mathbb{P}^1, \mathbb{P}^2)} \mathbb{E}^{\mathbb{P}}[\pi] \geq \sup_{\mathbb{P} \in \mathcal{M}(\mathbb{P}^1, \mathbb{P}^2)} \mathbb{E}^{\mathbb{P}}[c(S_1, S_2)1_A]$$

$$\geq \widetilde{\mathrm{MK}}_2^*(p)$$

By taking $\epsilon \to 0$, we get the reverse inequality $\widetilde{\mathrm{MK}}_2(p) \geq \widetilde{\mathrm{MK}}_2^*(p)$ from which we conclude. □

COROLLARY 2.3

For all $p > 0$,

$$\widetilde{\mathrm{MK}}_2(p) = \widetilde{\mathrm{MK}}_2(1) = \widetilde{\mathrm{MK}}_2$$

PROOF Note that for $p > 0$, $\inf_{\mathbb{P} \in \mathcal{P}(\mathbb{R}_+^2)} \mathbb{P}[A] = 0$ except if $A = \Omega$ for which $\inf_{\mathbb{P} \in \mathcal{P}(\mathbb{R}_+^2)} \mathbb{P}[A] = 1$. □

The robust quantile hedging price coincides therefore with our robust super-replication price. Nothing new arises by considering this robust quantile hedging. For completeness, we consider robust pricing using an utility preference function.

DEFINITION 2.13 *Let U be a strictly increasing concave function.*

$$\widetilde{\mathrm{MK}}_2(U, \alpha) \equiv \inf_{\mathcal{M}_p^*(\mathbb{P}^1, \mathbb{P}^2, U, \alpha)} \mathbb{E}^{\mathbb{P}^1}[\lambda_1(S_1)] + \mathbb{E}^{\mathbb{P}^2}[\lambda_2(S_2)] \qquad (2.52)$$

where $\mathcal{M}_p^*(\mathbb{P}^1, \mathbb{P}^2, U, \alpha)$ is the set of functions $\lambda_1 \in \mathrm{L}^1(\mathbb{P}^1), \lambda_2 \in \mathrm{L}^1(\mathbb{P}^2)$ and H a bounded continuous function on $\mathbb{R}_+$ such that

$$\inf_{\mathbb{P} \in \mathcal{P}(\mathbb{R}_+^2)} \mathbb{E}^{\mathbb{P}}[U(\lambda_1(S_1) + \lambda_2(S_2) + H(S_1)(S_2 - S_1))] \geq \alpha \qquad (2.53)$$

The following duality result shows that computing $\widetilde{\mathrm{MK}}_2(U, \alpha)$ is equivalent to solving our initial MOT $\widetilde{\mathrm{MK}}_2$.

THEOREM 2.13 see [12]
Assume that $\mathbb{P}^1 \leq \mathbb{P}^2$ are probability measures on $\mathbb{R}_+$ with first moments S_0 and Assumption 3 holds. Then the following duality holds:

$$\widetilde{\mathrm{MK}}_2(U, \alpha) = \widetilde{\mathrm{MK}}_2 + U^{-1}(\alpha)$$

2.2.11 Model-independent arbitrage

The classical fundamental theorem of asset pricing, briefly recapped in Theorem 1.2 in our discrete-time setup, forms the very basis of the rigorous mathematical theory of option pricing. The derivation of a comparable result in the model-independent setup is a central problem. It corresponds to the situation where the trader is allowed to trade Vanilla options. Even though there is a number of contributions in this direction ([2, 11, 13, 26, 51, 56] among others), the question to derive a completely satisfactory robust fundamental theorem remains open, mainly in the continuous-time setup. Here we briefly report some results.

DEFINITION 2.14 *There is a model-independent arbitrage if there exists a hedging strategy $(H_i)_{i=1,\ldots,n-1}$ bounded continuous and if there exists $(\lambda_i)_{i=1,\ldots,n} \in \mathrm{L}^1(\mathbb{P}^i)$ such that $\sum_{i=1}^n \mathbb{E}^{\mathbb{P}^i}[\lambda_i(S_i)] < 0$ and*

$$\sum_{i=1}^n \lambda_i(s_i) + \sum_{i=1}^{n-1} H_i(s_1, \ldots, s_i)(s_{i+1} - s_i) \geq 0$$

for all $(s_1, \ldots, s_n)$ in $\mathbb{R}_+^n$.

DEFINITION 2.15   *We say that C is a model-independent arbitrage-free price of a payoff $F_T = c(s_1, \ldots, s_n)$ if there does not exist a hedging strategy $(H_i)_{i=1,\ldots,n-1}$ bounded continuous and $(\lambda_i)_{i=1,\ldots,n} \in L^1(\mathbb{P}^i)$ for which $\sum_{i=1}^n \mathbb{E}^{\mathbb{P}^i}[\lambda_i(S_i)] < 0$ and*

$$\sum_{i=1}^n \lambda_i(s_i) + \sum_{i=1}^{n-1} H_i(s_1, \ldots, s_i)(s_{i+1} - s_i) + C - F_T \geq 0$$

for all $(s_1, \ldots, s_n)$ in $\mathbb{R}_+^n$.

Theorem 2.11 implies that

COROLLARY 2.4
There is no model-independent arbitrage if and only if $\mathcal{M}(\mathbb{P}^1, \ldots, \mathbb{P}^n) \neq \emptyset$.

Strassen's theorem [135] states that $\mathcal{M}(\mathbb{P}^1, \ldots, \mathbb{P}^n) \neq \emptyset$ if and only if $\mathbb{P}^1 \leq \mathbb{P}^2 \leq \ldots \leq \mathbb{P}^n$ in convex order. See also Corollary 4.1.

THEOREM 2.14
Let C be a model-independent arbitrage-free price. Then,
(i):

$$C_{\text{buy}} \equiv \inf_{\mathbb{Q} \in \mathcal{M}(\mathbb{P}^1, \ldots, \mathbb{P}^n)} \mathbb{E}^{\mathbb{Q}}[F_T] \leq C \leq C_{\text{sel}} \equiv \sup_{\mathbb{Q} \in \mathcal{M}(\mathbb{P}^1, \ldots, \mathbb{P}^n)} \mathbb{E}^{\mathbb{Q}}[F_T]$$

(ii): There exists $\mathbb{Q}^* \in \mathcal{M}(\mathbb{P}^1, \ldots, \mathbb{P}^n)$ such that

$$C = \mathbb{E}^{\mathbb{Q}^*}[F_T]$$

PROOF   **(i)** Similar proof as in Theorem 1.2.
(ii) Note that $\overline{\mathcal{M}}(\mathbb{P}^1, \ldots, \mathbb{P}^n) = \mathcal{M}(\mathbb{P}^1, \ldots, \mathbb{P}^n)$. We don't need to take the closure as for $\mathcal{M}_1$.   ▯

2.2.12   Market frictions

In practice, hedging Vanillas and delta-hedging induce transaction costs. They have been previously neglected. We show how they can be easily included in our framework. (i) There are bid-ask prices on Vanillas: this means that for Vanilla payoffs $(\lambda_i^\alpha(S_i))_{\alpha=1,\ldots,M}$ with maturity t_i, the market prices ranges in $[\underline{\Lambda}_i^\alpha, \overline{\Lambda}_i^\alpha]$, the buyer's (resp. seller's) price being $\underline{\Lambda}_i^\alpha$ (resp. $\overline{\Lambda}_i^\alpha$). (ii) Delta hedging at t_i, by changing the position in the underlying from H_{i-1} to H_i, induces a (proportional) transaction cost $\kappa|H_i - H_{i-1}|$ with $\kappa > 0$. The robust super-replication price can then be defined as

DEFINITION 2.16

$$\widetilde{\mathrm{MK}}_n^\kappa \equiv \inf_{\mathcal{M}^\kappa(\underline{\Lambda},\overline{\Lambda})} \sum_{\alpha=1}^M \sum_{i=1}^n \overline{\omega}_i^\alpha \overline{\Lambda}_i^\alpha - \underline{\omega}_i^\alpha \underline{\Lambda}_i^\alpha \tag{2.54}$$

where $\mathcal{M}^\kappa(\underline{\Lambda},\overline{\Lambda})$ is the set of $(\overline{\omega}_i^\alpha, \underline{\omega}_i^\alpha)_{1\le i\le n-1} \in \mathbb{R}_+$ and $(H_i)_{1\le i\le n-1}$ bounded continuous functions on $\mathbb{R}_+^i$ such that

$$\sum_{\alpha=1}^M \sum_{i=1}^n (\overline{\omega}_i^\alpha - \underline{\omega}_i^\alpha)\lambda_i^\alpha(s_i) + \sum_{i=1}^{n-1} H_i(s_1,\dots,s_i)(s_{i+1} - s_i)$$

$$+\kappa \sum_{i=0}^{n-1} |H_i - H_{i-1}| \ge c(s_1,\dots,s_n)$$

for all $(s_1,\dots,s_n) \in \mathbb{R}_+^n$. Here $H_{-1} = 0$ and $H_i \equiv H_i(s_1,\dots,s_i)$.

By using a minimax argument as in Theorem 2.1 (see [64, 71] for details), the dual formulation is

THEOREM 2.15

$$\widetilde{\mathrm{MK}}_n^\kappa = \sup_{\mathbb{P}\in\mathcal{M}^\kappa(\underline{\Lambda},\overline{\Lambda})} \mathbb{E}^{\mathbb{P}}[c(S_1,\dots,S_n)]$$

where

$$\mathcal{M}^\kappa(\underline{\Lambda},\overline{\Lambda}) = \{\mathbb{P} \: : \: \mathbb{E}^{\mathbb{P}}[\lambda_i^\alpha(S_i)] \in [\underline{\Lambda}_i^\alpha, \overline{\Lambda}_i^\alpha], \quad \mathbb{E}_{t_{i-1}}^{\mathbb{P}}[S_i - S_{i-1}] \in [-\kappa S_{i-1}, \kappa S_{i-1}],$$
$$\forall\, i = 1,\dots,n, \quad \forall\, \alpha = 1,\dots,M\}$$

2.3   Other optimal solutions

Dual side

As in Section 2.2.3, we are looking for an optimal martingale measure $\mathbb{P}^*$ supported on two maps T_d, T_u - minimal number of maps needed to ensure the martingale constraint:

$$\mathbb{P}^*(ds_1, ds_2) = \{q(s_1)\delta_{T_d(s_1)}(ds_2) + (1-q)(s_1)\delta_{T_u(s_1)}(ds_2)\}\, \mathbb{P}^1(ds_1) \tag{2.55}$$

In order to be feasible, i.e., $\mathbb{P}^* \in \mathcal{M}(\mathbb{P}^1, \mathbb{P}^2)$, $\mathbb{P}^*$ should satisfy:

$$q(x) = \frac{T_u(x) - x}{T_u(x) - T_d(x)} \tag{2.56}$$

and for $\psi \in L^1(\mathbb{P}^2)$

$$\int \psi(x)\mathbb{P}^2(dx) = \int \left(\psi(T_d(x))\frac{T_u(x) - x}{T_u(x) - T_d(x)} + \psi(T_u(x))\frac{x - T_d(x)}{T_u(x) - T_d(x)} \right)$$
$$\mathbb{P}^1(dx) \tag{2.57}$$

As $q \geq 0$, we should have $T_d(x) \leq x \leq T_u(x)$. Equation (2.56) corresponds to the martingale condition $\mathbb{E}[S_2|S_1] = S_1$ and Equation (2.57) to the marginal constraint $S_2 \overset{\mathbb{P}}{\sim} \mathbb{P}^2$.

REMARK 2.8   Equation (2.57) can be integrated out as (see [105] for a similar expression, Equation (12))

$$F_2(x) = F_1(T_u^{-1}(x)) + \int_{T_u^{-1}(x)}^{T_d^{-1}(x)} \frac{T_u(y) - y}{T_u(y) - T_d(y)}\mathbb{P}^1(dy) \tag{2.58}$$

taking T_d and T_u to be increasing functions. The first part of this equation, $F_2(x) = F_1(T_u^{-1}(x)) \cdots$, corresponds to the Fréchet–Hoeffding solution. In the case T_u increasing and T_d decreasing for $x \geq m$, we have our martingale Fréchet–Hoeffding solution (see Equations (2.31), (2.32)).   □

Primal side

We assume that there exists a primal maximizer to $\widetilde{MK}_2$ (note that this is not always the case - see an explicit counterexample in [17]), i.e., there exists $\lambda_1^* \in L^1(\mathbb{P}^1)$, $\lambda_2^* \in L^1(\mathbb{P}^2)$ and $H^* \in C_b(\mathbb{R}_+)$ such that

$$\lambda_1^*(s_1) + \lambda_2^*(s_2) + H^*(s_1)(s_2 - s_1) \geq c(s_1, s_2), \quad \forall (s_1, s_2) \in \mathbb{R}_+^2 \tag{2.59}$$

and $\widetilde{MK}_2 = \mathbb{E}^{\mathbb{P}^1}[\lambda_1^*] + \mathbb{E}^{\mathbb{P}^2}[\lambda_2^*]$. Below, for the simplicity of the notations, we disregard the superscript $*$. From (2.29), we have

$$\lambda_1(s_1) + \lambda_2(s_2) + H(s_1)(s_2 - s_1) = c(s_1, s_2), \quad \mathbb{P}^* - \text{a.s}$$

If $\mathbb{P}^*$ is of the form (2.55), this means that $\forall x \in \mathbb{R}_+$,

$$\lambda_1(x) + \lambda_2(T_d(x)) + H(x)(T_d(x) - x) = c(x, T_d(x)) \tag{2.60}$$
$$\lambda_1(x) + \lambda_2(T_u(x)) + H(x)(T_u(x) - x) = c(x, T_u(x)) \tag{2.61}$$

which is equivalent to

$$\lambda_1(x) \equiv \tag{2.62}$$
$$\frac{(c(x, T_d(x)) - \lambda_2(T_d(x)))(x - T_u(x)) - (c(x, T_u(x)) - \lambda_2(T_u(x)))(x - T_d(x))}{T_d(x) - T_u(x)}$$

$$H(x) = \frac{(c(x, T_d(x)) - \lambda_2(T_d(x))) - (c(x, T_u(x)) - \lambda_2(T_u(x)))}{T_d(x) - T_u(x)} \tag{2.63}$$

In particular, we assume that the infimum in (2.59) is reached for the two points $s_2 = T_d(s_1)$ and $s_2 = T_u(s_1)$ meaning that[6]

$$\lambda_2'(T_d(x)) + H(x) = c_2(x, T_d(x)) \tag{2.64}$$

$$\lambda_2'(T_u(x)) + H(x) = c_2(x, T_u(x)) \tag{2.65}$$

We have four equations (2.60), (2.61), (2.64), (2.65) for three unknown functions λ_1, λ_2 and h. By using our expression (2.63) for H, we get

$$\lambda_2'(T_d(x)) - \frac{\lambda_2(T_u(x)) - \lambda_2(T_d(x))}{T_u(x) - T_d(x)}$$

$$= c_2(x, T_d(x)) - \frac{c(x, T_u(x)) - c(x, T_d(x))}{T_u(x) - T_d(x)} \equiv \chi_1(x) \tag{2.66}$$

$$\lambda_2'(T_u(x)) - \frac{\lambda_2(T_u(x)) - \lambda_2(T_d(x))}{T_u(x) - T_d(x)}$$

$$= c_2(x, T_u(x)) - \frac{c(x, T_u(x)) - c(x, T_d(x))}{T_u(x) - T_d(x)} \equiv \chi_2(x) \tag{2.67}$$

By differentiating the equation (2.66) with respect to x and then using equation (2.67), we obtain

$$\lambda_2''(T_d)T_d' - \frac{T_d'}{T_u - T_d}\left(\lambda_2'(T_u) - \lambda_2'(T_d)\right) = \chi_1' + \chi_2 \frac{T_u' - T_d'}{T_u - T_d}$$

Finally, by subtracting Equation (2.67) to Equation (2.66), $\lambda_2'(T_u) - \lambda_2'(T_d) = \chi_2 - \chi_1$, we get

$$\frac{d}{dx}\left(\lambda_2'(T_d(x)) - c_2(x, T_d(x))\right) = -\frac{c_1(x, T_u(x)) - c_1(x, T_d(x))}{T_u(x) - T_d(x)}$$

By integrating this equation, we obtain for $x_0 \in \mathbb{R}_+$

$$\lambda_2'(T_d(x)) - \lambda_2'(T_d(x_0)) = c_2(x, T_d(x)) - c_2(x_0, T_d(x_0))$$

$$- \int_{x_0}^x \frac{c_1(y, T_u(y)) - c_1(y, T_d(y))}{T_u(y) - T_d(y)} dy \tag{2.68}$$

Finally, by subtracting Equation (2.64) to Equation (2.65), i.e., $\lambda_2'(T_d(x)) - \lambda_2'(T_u(x)) = c_2(x, T_d(x)) - c_2(x, T_u(x))$, we obtain

$$\lambda_2'(T_u(x)) - \lambda_2'(T_d(x_0)) = c_2(x, T_u(x)) - c_2(x_0, T_d(x_0))$$

$$- \int_{x_0}^x \frac{c_1(y, T_u(y)) - c_1(y, T_d(y))}{T_u(y) - T_d(y)} dy \tag{2.69}$$

---

[6]$c_2(x, y) = \partial_y c(x, y)$ (resp. $c_1(x, y) = \partial_x c(x, y)$).

This gives the *compatibility equation*:

$$c_2(T_d^{-1}(x), x) - c_2(T_u^{-1}(x), x) = \int_{T_u^{-1}(x)}^{T_d^{-1}(x)} \frac{c_1(y, T_u(y)) - c_1(y, T_d(y))}{T_u(y) - T_d(y)} dy$$

$$, \quad \forall\, x \in \mathrm{Dom}(T_d^{-1}) \bigcap \mathrm{Dom}(T_u^{-1}) \tag{2.70}$$

Below, we will check that: **(i)** under suitable conditions on the payoff c, the triplet $(\lambda_1, \lambda_2, H)$, as given explicitly by Equations (2.62, 2.63, 2.68, 2.69), defines a superhedging strategy. **(ii)** There exists a (unique) solution to equations (2.57)-(2.70). The solution to MOT is then completed from the weak duality result as in the proof of Theorem 2.8.

Condition on the payoff c

For each $s_1 \in \mathbb{R}_+$, the function $s_2 \mapsto c(s_1, s_2) - \lambda_2(s_2) - H(s_1)(s_2 - s_1)$ should admit two maximums at the points $s_2 = T_d(s_1)$ and $s_2 = T_u(s_1)$. The critical point equation is (with the compatibility equation (2.70))

$$c_2(s_1, s_2) - c_2(T_u^{-1}(s_2), s_2) = \int_{T_u^{-1}(s_2)}^{s_1} \frac{c_1(y, T_u(y)) - c_1(y, T_d(y))}{T_u(y) - T_d(y)} dy \tag{2.71}$$

We need to check that these critical points are maximum:

$$c_{22}(s_1, s_2) - \lambda_2''(s_2) \le 0 \text{ for } s_2 = T_i(s_1), \quad i = u, d$$

By using Equations (2.68), (2.69), this leads to

$$\left(c_{12}(s, T_u(s)) - \frac{c_1(s, T_u(s)) - c_1(s, T_d(s))}{T_u(s) - T_d(s)} \right) \frac{1}{T_u'(s)} \ge 0 \tag{2.72}$$

$$\left(c_{12}(s, T_d(s)) - \frac{c_1(s, T_u(s)) - c_1(s, T_d(s))}{T_u(s) - T_d(s)} \right) \frac{1}{T_d'(s)} \ge 0 \tag{2.73}$$

These two equations are satisfied for example : **(a)** if $c_{122} > 0$, T_u is increasing and T_d decreasing when $T_d \ne T_u$ or **(b)** T_u, T_d increasing and $c(s_1, s_2) = (s_2 - s_1)^+$ (or $c(s_1, s_2) = (s_2/s_1 - 1)^+$).
Finally, a straightforward computation proves that $(\lambda_1, \lambda_2, H)$ as defined above - everything is explicit - is in $\mathcal{M}^*(\mathbb{P}^1, \mathbb{P}^2)$ under condition **(a)** or **(b)**. This reproduces proof of Theorem 2.8 and we get additional optimal solutions for specific payoffs:

Example 2.10 Forward-start option[7], $c(s_1, s_2) = (s_2 - s_1)^+$
As a sanity check, we reproduce the upper bound as given in [105]:

$$\widetilde{\mathrm{MK}_2} = \int_0^\infty \frac{(T_u(x) - x)(x - T_d(x))}{T_u(x) - T_d(x)} \mathbb{P}^1(dx)$$

---

[7] Note that $|s_2 - s_1| = 2(s_2 - s_1)^+ - (s_2 - s_1)$ corresponds to the original Monge cost.

where T_d, T_u are two *increasing* functions satisfying (see Equations (2.58), (2.70))

$$F_2(x) = F_1(T_u^{-1}(x)) + \int_{T_u^{-1}(x)}^{T_d^{-1}(x)} \frac{T_u(y) - y}{T_u(y) - T_d(y)} \mathbb{P}^1(dy) \qquad (2.74)$$

and

$$\int_{T_u^{-1}(x)}^{T_d^{-1}(x)} \frac{1}{T_u(y) - T_d(y)} dy = 1$$

In particular, the maps (T_d, T_u) differ from our martingale Fréchet–Hoeffding solution. Finding numerically the two maps (T_d, T_u) by solving these highly nonlinear equations is a difficult task that we have not been able to succeed.

Note that the optimal lower bound is derived in [104] (see also [34]). Under Assumption 4(**Assum**($\mathbb{P}^1, \mathbb{P}^2$)), the optimal probability measure $\mathbb{P}^*$ is supported on *three* maps:

$$\mathbb{P}^*(ds_1, ds_2) = \mathbb{P}^1(ds_1)\left(\delta_{s_1}(ds_2)1_{s_1 \in (0,m] \cup [m^*, \infty)} + 1_{m < s_1 < m^*}\right.$$
$$\left.(q(s_1)\delta_{T_d(s_1)}(ds_2) + Q(s_1)\delta_{T_u(s_1)}(ds_2) + (1 - q(s_1) - Q(s_1))\delta_{s_1}(ds_2))\right)$$

where m^* (resp. m) is the global minimiser (resp. maximiser) of $\delta F \equiv F_2 - F_1$. $T_d : (m, m^*) \to [0, m]$ and $T_u : (m, m^*) \to [m^*, \infty)$ are continuous decreasing functions solution of

$$\delta F(T_u) + \delta F(T_d) = \delta F$$
$$\delta G(T_u) + \delta G(T_d) = \delta G$$

with $\delta G = G_2 - G_1$ and $G_i = \int_0^x y \mathbb{P}^i(dy)$. Then, $q, Q : (m, m^*) \to [0, 1]$ are given by

$$q(s) = \frac{s - T_d(s)}{T_u(s) - T_d(s)} \frac{f_1(s) - f_2(s)}{f_1(s)}$$
$$Q(s) = \frac{T_u(s) - s}{T_u(s) - T_d(s)} \frac{f_1(s) - f_2(s)}{f_1(s)}$$

where $\mathbb{P}^i(dx) \equiv f_i(x)dx$.

□

Example 2.11 Cliquet, $c(s_1, s_2) = \left(\frac{s_2}{s_1} - 1\right)^+$
The upper bound is

$$\widetilde{MK_2} = \int_0^\infty \frac{(T_u(x) - x)(x - T_d(x))}{x(T_u(x) - T_d(x))} \mathbb{P}^1(dx)$$

T_d, T_u are two *increasing* functions satisfying (2.74) and

$$\frac{1}{x} = \int_{T_u^{-1}(x)}^{T_d^{-1}(x)} \frac{T_u(y)}{y^2 \left(T_u(y) - T_d(y)\right)} dy \qquad (2.75)$$

□

2.4   Numerical experiments

We have seen that MOT can be solved under the martingale Spence–Mirrlees condition (see the previous section for other optimal solutions for some specific payoffs). In this section, we solve numerically MOT for various payoffs. The corresponding (finite-dimensional) linear programming is displayed in Example 2.6 for the payoff $(s_2/s_1 - K)^+$. For each payoff, we plot the optimal density and show that it is supported on some maps (see Figures 2.5, 2.6, 2.7). $\mathbb{P}^1$ (resp. $\mathbb{P}^2$) is given by a log-normal density with volatility $\sigma = 0.2$ and $t_1 = 1$ year (resp. $t_2 = 1.5$ years). For the payoff $c(s_1, s_2) = s_1(s_2 - 1)^2$, we reproduce numerically that the optimal measure is supported on two maps, one increasing and one decreasing when $x \geq m$ (see Figure 2.5) as the condition $c_{122} > 0$ is satisfied. Furthermore, the optimal measures for the two payoffs $c(s_1, s_2) = s_1(s_2 - 1)^3$, and $c(s_1, s_2) = s_1(s_2 - 1)^5$ are very close and seem to coincide (see Figures 2.6, 2.7).

2.5   Constrained MOT

Some variants in OT have been studied recently. In [112], the authors introduce an optimal transport with capacity constraints, which consists in minimising a cost among joint densities $\mathbb{P}$ with marginals $\mathbb{P}^1$ and $\mathbb{P}^2$ and under the *capacity* constraint

$$\mathbb{P}(s_1, s_2) \leq \bar{\mathbb{P}}(s_1, s_2)$$

for some prior joint density $\bar{\mathbb{P}}(s_1, s_2)$. In [37], optimal transports with congestion are considered. In this section, we present three variants of our MOT. The first one involves a constraint on the VIX future, the second one involves a penalty entropy term and the last one considers American options.

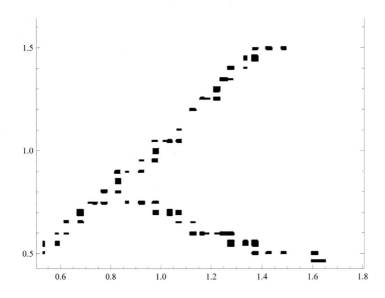

Figure 2.5: Optimal density for the payoff $c(s_1, s_2) = s_1(s_2 - 1)^2$.

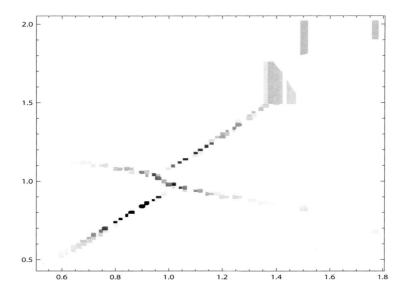

Figure 2.6: Optimal density for the payoff $c(s_1, s_2) = s_1(s_2 - 1)^3$.

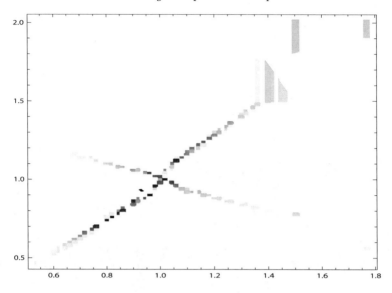

Figure 2.7: Optimal density for the payoff $c(s_1, s_2) = s_1(s_2 - 1)^5$.

2.5.1 VIX constraints

VIX futures and VIX options, traded on the CBOE, have become popular volatility derivatives. The payoff of a VIX index at a future expiry t_1 is by definition the price at t_1 of the 30 day log-contract which pays $-\frac{2}{t_2 - t_1} \ln \frac{S_2}{S_1}$ at $t_2 = t_1 + 30$ days:

$$\text{VIX}_{t_1}^2 \equiv -\frac{2}{\Delta} \mathbb{E}_{t_1}^{\mathbb{P}^{\text{mkt}}} \left[\ln \left(\frac{S_2}{S_1} \right) \right], \quad \Delta = t_2 - t_1 \tag{2.76}$$

This definition is at first sight strange as VIX_{t_1} seems to depend on the probability measure $\mathbb{P}^{\text{mkt}}$ (i.e., pricing model) used to value the log-contract at t_1. A choice should therefore be made and the probability measure $\mathbb{P}^{\text{mkt}}$ selected should be included in the term sheet which describes the payoff to the client. In fact, this conclusion is not correct and the value VIX_{t_1} is independent of the choice of $\mathbb{P}^{\text{mkt}}$ (i.e., model-independence). Indeed, from formula (2.2), the log-contract $\ln \frac{S_2}{S_1}$ with maturity t_2 can be replicated at t_1 with t_2-Vanillas:

$$\ln \frac{S_2}{S_1} = \frac{S_2 - S_1}{S_1} - \int_0^{S_1} \frac{dK}{K^2} (K - S_2)^+ dK - \int_{S_1}^{\infty} \frac{dK}{K^2} (S_2 - K)^+ dK$$

This implies that the arbitrage-free price (model-independent) at t_1 can be implied from the t_1 market value of t_2-Vanillas:

$$\text{VIX}^2_{t_1} \equiv \frac{2}{\Delta} \left(\int_0^{S_1} \frac{P(t_1, t_2, K)}{K^2} dK + \int_0^{S_1} \frac{C(t_1, t_2, K)}{K^2} dK \right)$$

with $P(t_1, t_2, K)$ (resp. $C(t_1, t_2, K)$) the undiscounted market price at t_1 of a put (resp. call) option with strike K and maturity t_2.

The payoff of a call option on VIX expiring at t_1 with strike K is $(\text{VIX}_{t_1} - K)^+$. Below, the market value (at $t = 0$) for the VIX future (i.e., $K = 0$) is denoted VIX.

Assumption 5 *For technical reason, we will assume that the random variables $(S_{t_1}, S_{t_2}, \text{VIX}_{t_1})$ are supported on a compact interval $I_1 \times I_2 \times I_X \subset (\mathbb{R}^*_+)^3$.*

For further reference, we denote by $\mathcal{M}(\mathbb{P}^1, \mathbb{P}^2, \text{VIX})$ the set of all martingale measures $\mathbb{P}$ on the (pathspace) $I_1 \times I_2 \times I_X$ having marginals $\mathbb{P}^1$, $\mathbb{P}^2$ with mean S_0 and such that $\text{VIX} = \mathbb{E}^\mathbb{P}[\text{VIX}_{t_1}]$, that is:

$$\mathcal{M}(\mathbb{P}^1, \mathbb{P}^2, \text{VIX}) = \Big\{ \mathbb{P} \in \mathcal{P}(I_1 \times I_2 \times I_X) : \ S_1 \overset{\mathbb{P}}{\sim} \mathbb{P}^1, \ S_2 \overset{\mathbb{P}}{\sim} \mathbb{P}^2,$$

$$\mathbb{E}^\mathbb{P}[\text{VIX}_{t_1}] = \text{VIX},$$

$$\mathbb{E}^\mathbb{P}[S_2 | S_1, \text{VIX}_{t_1}] = S_1, \tag{2.77}$$

$$\mathbb{E}^\mathbb{P}\left[-\frac{2}{\Delta} \log \frac{S_2}{S_1} | S_1, \text{VIX}_{t_1} \right] = \text{VIX}^2_{t_1} \Big\}$$

We define our constrained MOT for a VIX call option expiring at t_1 with strike K as

DEFINITION 2.17

$$\text{MK}_{\text{vix}} \equiv \inf_{\lambda_1 \in L^1(\mathbb{P}^1), \lambda_2 \in L^1(\mathbb{P}^2), \lambda \in \mathbb{R}, H_S, H_X} \mathbb{E}^{\mathbb{P}^1}[\lambda_1(S_1)] + \mathbb{E}^{\mathbb{P}^2}[\lambda_2(S_2)] + \lambda \text{VIX}$$

such that for all $(s_1, s_2, x) \in I_1 \times I_2 \times I_X$,

$$\lambda_1(s_1) + \lambda_2(s_2) + \lambda\sqrt{x} + H_S(s_1, x)(s_2 - s_1) \tag{2.78}$$

$$+ H_X(s_1, x)\left(-\frac{2}{\Delta} \ln\left(\frac{s_2}{s_1}\right) - x \right) \geq (\sqrt{x} - K)^+$$

where the functions $H_S, H_X : I_1 \times I_X \to \mathbb{R}$ are assumed to be bounded continuous functions on $I_1 \times I_X$, $\lambda_1 \in L^1(\mathbb{P}^1)$ and $\lambda_2 \in L^1(\mathbb{P}^2)$. In [88], sharp bounds for the prices of VIX futures alone are derived.

Note that this defines a *linear* semi-infinite infinite-dimensional programming problem. The variable x should be interpreted as the t_1-value of a log-contract

$-2/\Delta \ln \frac{s_2}{s_1}$, i.e., the square of the VIX index $\mathrm{VIX}_{t_1}^2$. This semi-static super-replication consists in holding statically t_1 and t_2-European payoffs with market prices $\mathbb{E}^{\mathbb{P}^1}[\lambda_1(S_1)]$ and $\mathbb{E}^{\mathbb{P}^2}[\lambda_2(S_2)]$, a VIX future with market price VIX and delta hedging at t_1 (with zero-cost) on the spot and on a forward log-contract with price x. The value of this portfolio is greater than or equal to the payoff $(\sqrt{x} - K)^+$. If somebody offers this VIX option at a price p above $\mathrm{MK}_{\mathrm{vix}}$, the arbitrage can be locked in by selling this option and going long in the above super-replication:

$$\lambda_1(s_1) + \lambda_2(s_2) + \lambda\sqrt{x} + H_S(s_1, x)(s_2 - s_1) + H_X(s_1, x)\left(-\frac{2}{\Delta}\ln\left(\frac{s_2}{s_1}\right) - x\right)$$

$$- \left(\sqrt{x} - K\right)^+ + (p - \mathrm{MK}_{\mathrm{vix}}) \geq 0, \quad \forall\, (s_1, s_2, x) \in I_1 \times I_2 \times I_X$$

We have a dual version which is connected to a constrained martingale optimal transport problem:

THEOREM 2.16 Duality, see [58] for a detailed proof
Assume that $\mathbb{P}^1$, $\mathbb{P}^2$ are probability measures respectively on I_1 and I_2 such that $\mathcal{M}(\mathbb{P}^1, \mathbb{P}^2, \mathrm{VIX})$ is non-empty. Then,

$$\mathrm{MK}_{\mathrm{vix}} = \max_{\mathbb{P} \in \mathcal{M}(\mathbb{P}^1, \mathbb{P}^2, \mathrm{VIX})} \mathbb{E}[(\mathrm{VIX}_{t_1} - K)^+]$$

This is part of the result that we have a max and not only a sup, meaning that the seller's price is attained by a martingale measure, i.e., a model calibrated to the t_1 and t_2 Vanillas and to the VIX future. In [88] (see Theorem 4.1), a similar theorem is proved where the Assumption 5 is relaxed.

PROOF (Sketch)   The function H_s is interpreted as the Lagrange multiplier associated to the martingale condition, the function H_X is associated to the constraint $\mathrm{VIX}_{t_1}^2 = \mathbb{E}_{t_1}^{\mathbb{P}}[-(2/\Delta)\ln S_2/S_1]$, the functions λ_1 and λ_2 to the marginal constraints and finally λ to the VIX constraint $\mathrm{VIX} = \mathbb{E}^{\mathbb{P}}[\mathrm{VIX}_{t_1}]$. Then, the proof copycats the one in Theorem 2.1. ∎

The dual corresponds to the maximization of the expectation of a VIX payoff with respect to a martingale measure with marginals $\mathbb{P}^1$, $\mathbb{P}^2$ and with the constraint on the VIX future $\mathrm{VIX} = \mathbb{E}^{\mathbb{P}}[\sqrt{\mathbb{E}_{t_1}^{\mathbb{P}}[-(2/\Delta)\ln S_2/S_1]}]$. Note that as this additional constraint, not present in the original MOT, is nonlinear with respect to the (martingale) measure $\mathbb{P}$, this OT problem is more involved. As a crucial step, by introducing a delta-hedging on a forward log-contract $-2/\Delta \ln s_2/s_2$, this problem has been converted into a linear programming problem (see Definition 2.17) that can be solved with a simplex algorithm. Note that a similar trick is used in [25] for converting a quantile hedging approach into a super-replication problem.

Optimality

Under technical conditions on the marginal $\mathbb{P}^2$, this MOT can be solved explicitly (see [58] for details).

First, one can prove that $\mathrm{MK}_{\mathrm{vix}} \leq \overline{\mathrm{MK}}_{\mathrm{vix}}$ where

$$\overline{\mathrm{MK}}_{\mathrm{vix}} = \begin{cases} \frac{1}{2}\left(\mathrm{VIX} - K + I\right) & K \geq K^* \equiv \frac{\sigma_{1,2}^2}{2\mathrm{VIX}} \\ \mathrm{VIX} - K\frac{\mathrm{VIX}^2}{\sigma_{1,2}^2} & K < K^* \end{cases} \qquad (2.79)$$

with $\sigma_{1,2}^2 \equiv -\frac{2}{\Delta}(\mathbb{E}^{\mathbb{P}^2}[\ln S_2] - \mathbb{E}^{\mathbb{P}^1}[\ln S_1])$. Indeed this upper bound $\overline{\mathrm{MK}}_{\mathrm{vix}}$ is attained by the semi-static super-replication

$$\lambda_1(s_1) = -\frac{2}{\Delta}H_X \ln\frac{s_1}{S_0} + \nu, \quad \lambda_2(s_2) = \frac{2}{\Delta}H_X \ln\frac{s_2}{S_0}, \quad H_S = 0 \quad (2.80)$$

where for $K \geq K^*$:

$$H_X = -\frac{1}{4I(K)}, \quad \nu = \frac{-2KI(K) + \sigma_{1,2}^2 - 2K(\mathrm{VIX} - K)}{4I(K)}, \quad \lambda = \frac{1}{2} - \frac{K}{2I(K)}$$

with $I(K) \equiv \sqrt{\sigma_{1,2}^2 - \mathrm{VIX}^2 + (\mathrm{VIX} - K)^2}$ and for $K \leq K^*$:

$$H_X = -K\left(\frac{\mathrm{VIX}}{\sigma_{1,2}^2}\right)^2, \quad \nu = 0, \quad \lambda = 1 - 2K\frac{\mathrm{VIX}}{\sigma_{1,2}^2}$$

For use below, we define a bi-atomic measure

$$\mathbb{P}(dx) = p_0\delta_{x_0}(dx) + p_1\delta_{x_1}(dx), \quad p_0 = p, \quad p_1 = (1 - p) \qquad (2.81)$$

with

$$\begin{cases} x_0 = K - I(K); \; x_1 = K + I(K); \; p = \frac{K - \mathrm{VIX} + I(K)}{2I(K)} & \text{if } K \geq K^*, \\ x_0 = 0; \qquad\qquad x_1 = \frac{\sigma_{1,2}^2}{\mathrm{VIX}}; \qquad p = \frac{\sigma_{1,2}^2 - \mathrm{VIX}^2}{\sigma_{1,2}^2} & \text{if } K < K^*, \end{cases}$$

Finally, under appropriate conditions on $(\mathbb{P}^1, \mathbb{P}^2)$, we have that $\overline{\mathrm{MK}}_{\mathrm{vix}}$ is the optimal bound:

THEOREM 2.17 see [58] for the proof
The following are equivalent:
(i): $\mathrm{MK}_{\mathrm{vix}} = \overline{\mathrm{MK}}_{\mathrm{vix}}$
(ii): *There exist two couples of measures (μ_0, ν_0) and (μ_1, ν_1) on $\mathbb{R}_+$ such that*

$$\mathbb{P}^1 = p\mu_0 + (1 - p)\mu_1, \qquad \mathbb{P}^2 = p\nu_0 + (1 - p)\nu_1, \qquad (2.82)$$

and

$$\int f\left(z, \log(z)\right) \nu_i(dz) \geq \int f\left(y, \log(y) - \frac{\Delta}{2} x_i^2\right) \mu_i(dy), \qquad i = 0, 1 \quad (2.83)$$

for every convex function $f : \mathbb{R}^2 \to \mathbb{R}$.

Note that checking (numerically) the condition (2.83) is not obvious. As an example of $(\mathbb{P}^1, \mathbb{P}^2)$ satisfying the condition **(ii)**, we take a log-normal distribution with mean 1 and volatility 0.4 (maturity = 1 year) for $\mathbb{P}^1$ and $\mathbb{P}^2$ is given by

$$\mathbb{P}^2(s) = \sum_{i=0}^{1} p_i \left(\mathbb{P}^1 \left(\frac{s}{\alpha_d^i} \right) \frac{\alpha_u^i - 1}{(\alpha_u^i - \alpha_d^i)\alpha_d^i} + \mathbb{P}^1 \left(\frac{s}{\alpha_u^i} \right) \frac{1 - \alpha_d^i}{(\alpha_u^i - \alpha_d^i)\alpha_u^i} \right)$$

with $\alpha_u^i > 1$ and $\alpha_d^i < 1$ uniquely fixed by

$$\frac{\alpha_u^i - 1}{\alpha_u^i - \alpha_d^i} \ln \alpha_d^i + \frac{1 - \alpha_d^i}{\alpha_u^i - \alpha_d^i} \ln \alpha_u^i = -\Delta \frac{x_i^2}{2}, \quad i = 0, 1$$

We have chosen VIX $= 0.3$, $\sigma_{1,2} = 0.4$. For $K = $ VIX and for the upper bound, we have $x_0 = 0.035$, $x_1 = 0.567$, $p = 0.5$. We have chosen $\alpha_u^0 = 1.1$, $\alpha_u^1 = 1.2$ ($\alpha_d^0 = 0.998$, $\alpha_d^1 = 0.863$ here). The densities $\mathbb{P}^1$ and $\mathbb{P}^2$ are plotted in Figure 2.8.

We have compared the analytical upper bound $\overline{\text{MK}}_{\text{vix}}$ against market prices for VIX options with expiry 16 Oct. 13, pricing date = 12 Jul. 2013 (see Figure 2.9). Surprisingly, market prices of VIX options are well-above $\overline{\text{MK}}_{\text{vix}}$, highlighting a market arbitrage. In order to see the impact of transaction cost on Vanillas, we have added $+0.5\%$ and $+1.0\%$ to the market value of the log-swap forward $\sigma_{1.2} = 18.15\%$ (VIX $= 18.05\%$ here). As the transaction cost increases, the arbitrage opportunities evaporate.

2.5.2   Entropy penalty

The main drawback of the (robust) super-replication approach, appearing in the definition of the MOT, is that the martingale measure (i.e., arbitrage-free model), which achieves the super-replication strategy, can be very different from those generated by (stochastic volatility) diffusive models traders commonly use. As an example, the upper bound given marginals (such that the associated barycenter functions are increasing in time) for an increasing payoff on the maximum of a martingale is reached by Azéma–Yor solution to SEP (see Section 4.6.1). This martingale belongs to the class of local jump Lévy models (see Section 4.2). In the following, we show that this drawback can be circumvented. To this end, we introduce the Kullback–Leibler relative

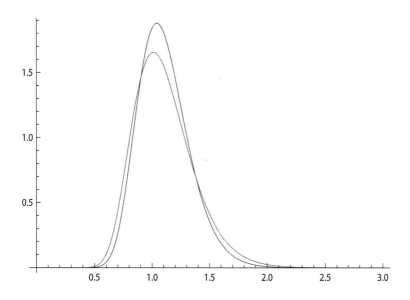

Figure 2.8: Example of $(\mathbb{P}^1, \mathbb{P}^2)$. We have chosen $\alpha_u^0 = 1.1$, $\alpha_u^1 = 1.2$ ($\alpha_d^0 = 0.998$, $\alpha_d^1 = 0.863$). The top (resp. low) curve is $\mathbb{P}^1$ (resp. $\mathbb{P}^2$).

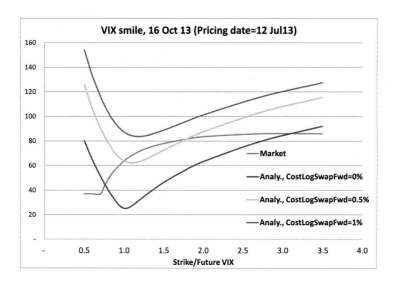

Figure 2.9: Analytical upper bounds versus market values for VIX smile t_1 =16 Oct. 13 (pricing date = 12 Jul. 2013). $\sigma_{1,2} = 18.15\%$ (VIX = 18.05%).

entropy between two probability measures $\mathbb{P}$ and $\mathbb{P}^0$:

$$H(\mathbb{P}, \mathbb{P}^0) = \mathbb{E}^{\mathbb{P}}\left[\ln \frac{d\mathbb{P}}{d\mathbb{P}^0}\right], \quad \mathbb{P} \text{ is absolutely continuous w.r.t. } \mathbb{P}^0$$
$$= +\infty, \quad \text{otherwise}$$

Our construction can also be generalized to other functional $\mathcal{F}(\mathbb{P}|\mathbb{P}^0)$ - see Example (2.17).
As

$$H(\mathbb{P}, \mathbb{P}^0) = \mathbb{E}^{\mathbb{P}^0}\left[\frac{d\mathbb{P}}{d\mathbb{P}^0} \ln \frac{d\mathbb{P}}{d\mathbb{P}^0}\right] = \mathbb{E}^{\mathbb{P}^0}\left[f \ln f - f + 1\right], \quad f \equiv \frac{d\mathbb{P}}{d\mathbb{P}^0}$$

we have that $H(\mathbb{P}, \mathbb{P}^0) \geq 0$.

Following the same route as for the robust super-replication $\widetilde{\mathrm{MK}}_c$, we impose that our model matches market marginals at each date $(t_i)_{i=1,\ldots,n}$ (eventually additional market instruments $(c_a)_{a=1,\ldots,N}$) and define the primal formulation:

DEFINITION 2.18 *Fix $\lambda \in \mathbb{R}_+$.*

$$\widetilde{\mathrm{MK}}_n^{\lambda} \equiv \sup\{\mathbb{E}^{\mathbb{P}}[c(S_1, \ldots, S_n)] : \mathbb{P} \in \mathcal{M}(\mathbb{P}^1, \ldots, \mathbb{P}^n), \quad H(\mathbb{P}, \mathbb{P}^0) \leq \lambda\}$$
$$(2.84)$$

We have added the constraint $H(\mathbb{P}, \mathbb{P}^0) \leq \lambda$ with λ a positive parameter: $\mathbb{P}$ should be "close" to $\mathbb{P}^0$. Note that $H(\mathbb{P}, \mathbb{P}^0)$ is not a distance. In this respect, one can use instead the Wasserstein distance which corresponds to MOT with a quadratic cost. Endowed with this Wasserstein distance, the space of probability measures becomes metrizable (see [139], Chapter 7).

By setting $\lambda = 0$, the feasible set is empty except if the prior measure $\mathbb{P}^0$ is a martingale measure and satisfies the marginal constraints (as $H(\mathbb{P}, \mathbb{P}^0) = 0$ if and only if $\mathbb{P} = \mathbb{P}^0$). In this case, we get

$$\widetilde{\mathrm{MK}}_n^0 = \mathbb{E}^{\mathbb{P}^0}[c]$$

By taking the limit $\lambda = \infty$, we obtain our initial MOT:

$$\widetilde{\mathrm{MK}}_n^{\infty} = \widetilde{\mathrm{MK}}_n$$

The parameter λ provides therefore an interpolation between prices given either by the prior (risk-neutral) model or the (robust) super-replication strategy.

Such an approach was explored in [4], [5], [6], [7]. The martingale constraints seem to have been unnoticed, in particular in [5] and [7], resulting in pricing models that are not arbitrage-free. We would like to emphasize that the density is not risk-neutral as long as the martingale constraint is not fulfilled. This is confirmed in Proposition 4 in [6], where the drift of the diffusion measure $\mathbb{P}^*$ is computed and found to be different from the risk-free interest rate.

The dual formulation of $\widetilde{\mathrm{MK}}_n^\lambda$ reads

DEFINITION 2.19

$$
\widetilde{\mathrm{MK}}_n^{\lambda\,*} \equiv \inf_{(\lambda_i(\cdot))_{1\leq i\leq n},(H_i(\cdot))_{1\leq i\leq n-1},\zeta\in\mathbb{R}_+} \sum_{i=1}^n \mathbb{E}^{\mathbb{P}^i}[\lambda_i(S_i)]
$$
$$
+\zeta\left(\lambda + \ln\mathbb{E}^{\mathbb{P}^0}\left[e^{\zeta^{-1}\left(c-\sum_{i=1}^n \lambda_i(S_i)-\sum_{i=1}^{n-1} H_i(S_1,\dots,S_i)(S_{i+1}-S_i)\right)}\right]\right)
$$

$$(2.85)$$

Here the Lagrange multiplier $\zeta \in \mathbb{R}_+$ corresponds to the additional constraint $H(\mathbb{P},\mathbb{P}^0) \leq \lambda$.

THEOREM 2.18 see [93]

Assume that the set $\{\mathbb{P}\in\mathcal{M}(\mathbb{P}^1,\dots,\mathbb{P}^n), H(\mathbb{P},\mathbb{P}^0)\leq\lambda\}$ is non-empty. Let $c:\mathbb{R}_+^n \to [0,\infty)$ be a continuous function so that

$$
c(s_1,\dots,s_n) \leq K\cdot(1+s_1+\dots+s_n)
$$

$$(2.86)$$

on $\mathbb{R}_+^n$ for some constant K. Then there is no duality gap $\widetilde{\mathrm{MK}}_n^{\lambda\,*} = \widetilde{\mathrm{MK}}_n^\lambda$. The supremum is attained by the optimal measure $\mathbb{P}^*$ given by the Gibbs density

$$
\frac{d\mathbb{P}^*}{d\mathbb{P}^0} = \frac{e^{(\zeta^*)^{-1}\left(c-\sum_{i=1}^n \lambda_i^*(S_i)-\sum_{i=1}^{n-1} H_i^*(S_{i+1}-S_i)\right)}}{\mathbb{E}^{\mathbb{P}^0}\left[e^{(\zeta^*)^{-1}\left(c-\sum_{i=1}^n \lambda_i^*(S_i)-\sum_{i=1}^{n-1} H_i^*(S_{i+1}-S_i)\right)}\right]}
$$

$$(2.87)$$

where $((\lambda_i^*(\cdot))_{1\leq i\leq n},(H_i^*(\cdot))_{1\leq i\leq n-1},\zeta^*)$ achieves the infimum in (2.85).

The sagacious reader will note that a similar duality result is used for proving Talagrand's inequalities in OT (see Chapter 9 in [139]). Also the relative entropy can be replaced by any convex functional $\mathcal{F}(\mathbb{P})$ (see some examples in Table 2.1).

An alternative approach, to take care of the martingale constraint, is to consider a random mixture of N-reference models where the distribution of mixture weights is obtained by solving a well-posed convex optimization problem (see [49]). This corresponds to a classical OT and the dual is connected to an exponential hedging formulation (see Theorem 3.5 in [49]).

PROOF (Sketch) By introducing Lagrange's multipliers, $\widetilde{\mathrm{MK}}_n^\lambda$ can be written as

$$\widetilde{\mathrm{MK}}_n^\lambda = \sup_{\mathbb{P} \in \mathcal{M}_+} \inf_{(\lambda_i(\cdot))_{1 \le i \le n}, (H_i(\cdot))_{1 \le i \le n-1}, \zeta \in \mathbb{R}_+} \mathbb{E}^{\mathbb{P}}[c] + \zeta\left(\lambda - H(\mathbb{P}, \mathbb{P}^0)\right)$$

$$+ \sum_{i=1}^{n}\left(\mathbb{E}^{\mathbb{P}^i}[\lambda_i(S_i)] - \mathbb{E}^{\mathbb{P}}[\lambda_i(S_i)]\right) - \sum_{i=1}^{n-1}\mathbb{E}^{\mathbb{P}}[H_i(S_{i+1} - S_i)]$$

where H_i is an arbitrary function of $(S_1, \ldots, S_i)$ and $\mathcal{M}_+$ is the set of positive measure on $\mathbb{R}_+^n$. By applying a minimax argument, we can permute the inf and the sup operators. Finally, the supremum over $\mathbb{P} \in \mathcal{M}_+$ is reached by the Gibbs density (2.87). Plugging this density above gives our dual formulation $\widetilde{\mathrm{MK}}_n^{\lambda*}$.   ▯

The probability measure $\mathbb{P}^*$ which achieves the maximum is payoff-dependent (see Equation (2.87) where $\mathbb{P}^*$ depends on c). In order to get a payoff-independent measure, we could introduce a new primal problem:

$$\mathrm{P}^0 \equiv \inf\{H(\mathbb{P}, \mathbb{P}^0) : \mathbb{P} \in \mathcal{M}(\mathbb{P}^1, \ldots, \mathbb{P}^n)\}$$

A variant of this primal, without the connections to model calibration, has been studied by many authors in the context of optimal portfolio choice under exponential utility (see for instance [60, 70]).

2.5.3   American options

We consider a two-period Bermudan option in which the holder has the right but not the obligation to exercise a payoff $g(S)$ at t_1 or t_2. The MOT problem is defined as

DEFINITION 2.20

$$\mathrm{MK}_{\mathrm{Ame}}(\mathbb{P}^1, \mathbb{P}^2) \equiv \inf_{\lambda_1, \lambda_2, H_1, H_2} \mathbb{E}^{\mathbb{P}^1}[\lambda_1(S_1)] + \mathbb{E}^{\mathbb{P}^2}[\lambda_2(S_2)]$$

such that

$$\lambda_1(s_1) + \lambda_2(s_2) + H_1(s_1)(s_2 - s_1) \ge g(s_1), \quad \forall\, (s_1, s_2) \in \mathbb{R}_+^2 : \text{exercise at } t_1$$
$$\lambda_1(s_1) + \lambda_2(s_2) + H_2(s_1)(s_2 - s_1) \ge g(s_2), \quad \forall\, (s_1, s_2) \in \mathbb{R}_+^2 : \text{exercise at } t_2$$

Here we have two delta functions $H_1(s_1)$ and $H_2(s_1)$ because our delta-hedging strategy depends at t_1 on the value S_{t_1} but also if the payoff g has been exercised or not. The dual is then

$$\mathrm{MK}_{\mathrm{Ame}}(\mathbb{P}^1, \mathbb{P}^2) = \sup_{p_1(s_1, s_2) \ge 0, p_2(s_1, s_2) \ge 0} \int g(s_1)p_1(ds_1, ds_2)$$

$$+ \int g(s_2)p_2(ds_1, ds_2)$$

such that

$$\int (s_2 - s_1)p_1(s_1, s_2)ds_2 = 0, \quad \int (s_2 - s_1)p_2(s_1, s_2)ds_2 = 0,$$

$$\int (p_1(s_1, s_2) + p_2(s_1, s_2))ds_2 = \mathbb{P}^1(s_1),$$

$$\int (p_1(s_1, s_2) + p_2(s_1, s_2))ds_1 = \mathbb{P}^2(s_2)$$

In the particular case where g is convex, the optimal primal is attained by

$$\lambda_1^*(s_1) = 0, \lambda_2^*(s_2) = g(s_2), H_1^*(s_1) = -g'(s_1), H_2^*(s_1) = 0$$

The optimal dual is reached by taking $p_1(ds_1, ds_2) = 0$ and p_2 an arbitrary martingale measure with marginals $\mathbb{P}^1$ and $\mathbb{P}^2$. Indeed, we have $\mathrm{MK}_{\mathrm{Ame}} \leq \mathbb{E}^{\mathbb{P}^2}[g(S_2)]$ and we deduce by weak duality the optimality of our guess.
The duality is studied in full generality in [10, 61, 106].
An open problem is the following: Apart from this trivial case, could this MOT be solved explicitly?

Chapter 3

Model-independent options

Abstract   We consider path-dependent payoffs for which seller and buyer's super-replication prices coincide. They can be perfectly replicated by dynamically hedging the underlying (and eventually static hedging in Vanillas). This class can be enlarged by restricting the class of risk-neutral models to be Ocone's martingales. An other class of model-independent payoffs is provided by timer-like options.

3.1   Probabilistic setup

This chapter requires some basic knowledge in stochastic analysis (not so much, mainly stochastic integration and Itô's formula).

As in Chapter 2, we assume a zero interest rate (non-zero rates are briefly considered in Section 3.3.4). The price of an asset at time t, $(S_t)_{t\in[0,T]}$, will be modeled by a continuous semi-martingales. The semi-martingale property is imposed as we want to give a meaning to the limit when $n \to \infty$ of the discrete delta-hedging

$$\sum_{t_0=0}^{T} H_{t_i}\left(S_{t_{i+1}} - S_{t_i}\right) \overset{n\to\infty}{\longrightarrow} \int_0^T H_t dS_t$$

Good integrator processes are precisely provided by semi-martingales. Below, we describe our probabilistic framework.

Let $\Omega \equiv \{\omega \in C([0,T], \mathbb{R}_+) : \omega_0 = 0\}$ be the canonical space equipped with the uniform norm $\|\omega\|_\infty \equiv \sup_{0\leq t\leq T} |\omega(t)|$, B the canonical process, i.e., $B_t(\omega) \equiv \omega(t)$ and $\mathcal{F} \equiv \{\mathcal{F}_t\}_{0\leq t\leq T}$ the filtration generated by B: $\mathcal{F}_t = \sigma\{B_s, s \leq t\}$. $\mathbb{P}^0$ is the Wiener measure. S_0 is some given initial value in $\mathbb{R}_+$, and we denote

$$S_t \equiv S_0 + B_t \text{ for } t \in [0, T].$$

For any $\mathcal{F}$−adapted process σ and satisfying $\int_0^T \sigma_s^2 ds < \infty$, $\mathbb{P}^0$−a.s., we define the probability measure on $(\Omega, \mathcal{F})$:

$$\mathbb{P}^\sigma \equiv \mathbb{P}^0 \circ (S^\sigma)^{-1} \text{ where } S_t^\sigma \equiv S_0 + \int_0^t \sigma_r dB_r, \ t \in [0, T], \ \mathbb{P}^0 - \text{a.s.}$$

Then S is a $\mathbb{P}^\sigma$−local martingale. We denote by $\mathcal{M}^c$ the collection of all such probability martingale measures on $(\Omega, \mathcal{F})$. The quadratic variation process $\langle S \rangle = \langle B \rangle$ takes values in the set of all nondecreasing continuous functions. Note that the quadratic variation can be defined pathwise as the limsup of the corresponding discrete counterpart with conveniently chosen mesh of the time partition. The dependence of the quadratic variation on the underlying probability measure $\mathbb{P} \in \mathcal{M}^c$ can therefore be dropped. Finally, $\mathcal{M}^c(\mu) \equiv \{\mathbb{P}^\sigma \in \mathcal{M}^c \ : \ S_T^\sigma \overset{\mathbb{P}^\sigma}{\sim} \mu\}$ where μ is supported on $\mathbb{R}_+$. For the ease of notation, we will delete the superscript σ on S^σ below.

Extremal points of $\mathcal{M}^c$ and complete models

$\mathcal{M}^c$ is a convex set (unfortunately not compact) and it is interesting to characterize its extremal points (see a previous similar Theorem 2.6).

> **THEOREM 3.1 [109], see also [130]**
> Let $\mathbb{P} \in \mathcal{M}^c$. The following two properties are equivalent:
>
>   1. $\mathbb{P} \in \mathrm{Ext}(\mathcal{M}^c)$.
>
>   2. *All* $\mathbb{P}$*-martingale* M_t *can be written as*
>
> $$M_t = C + \int_0^t H_s dS_s \tag{3.1}$$
>
> *where* $C \in \mathbb{R}$ *and* H *is a predictable process with respect to* $\mathcal{F}$*, i.e.,* $(s, \omega) \mapsto H_s(\omega)$ *is measurable with respect to the smallest* σ*-algebra on* $\mathbb{R}_+ \times \Omega$ *making all continuous and adapted processes measurable. In short, they are the processes that can be predicted from the strict knowledge of the past.*

This theorem shows that elements of $\mathrm{Ext}(\mathcal{M}^c)$ correspond precisely to complete models: The equation (3.1) indicates that the arbitrage-free price of an option, $M_t \equiv \mathbb{E}^{\mathbb{P}}[F_T | \mathcal{F}_t]$ which is a $\mathbb{P}$-martingale, can be dynamically replicated. In particular for $t = T$, we get that the payoff F_T is attainable:

$$F_T = \mathbb{E}^{\mathbb{P}}[F_T] + \int_0^T H_s dS_s \tag{3.2}$$

3.2   Exotic options made of Vanillas: Nice martingales

3.2.1   Variance swaps

It is well-known that the process $\ln S_t + \frac{1}{2}\langle \ln S \rangle_t$ is a martingale. As an important consequence in finance, this leads to the exact replication of a variance swap (within the class $\mathcal{M}^c$) in terms of a log-contract. A discrete-monitoring variance swap pays at a maturity T the sum of daily squared log-returns, mainly

$$\frac{1}{T} \sum_{i=0}^{n-1} \left(\ln \frac{S_{t_{i+1}}}{S_{t_i}} \right)^2, \quad t_0 = 0, \quad t_n = T$$

and $\Delta t = t_{i+1} - t_i =$ one day. In the limit $n \to \infty$, it converges $\mathbb{P}$-almost surely to the quadratic variation $\langle \ln S \rangle_T$ of $\ln S$:

$$\frac{1}{T} \sum_{i=0}^{n-1} \left(\ln \frac{S_{t_{i+1}}}{S_{t_i}} \right)^2 \xrightarrow{n \to \infty} \frac{1}{T} \langle \ln S \rangle_T$$

REMARK 3.1   Note that in practice, $t_{k+1} - t_k = 1$ day and the approximation of a discrete-monitored variance swap by its continuous-time version is valid. Indeed,

$$\text{VS} \equiv \frac{1}{T} \sum_{i=0}^{n-1} \mathbb{E}\left[\left(\ln \frac{S_{t_{i+1}}}{S_{t_i}} \right)^2 \right] = \frac{1}{T} \sum_{i=0}^{n-1} \mathbb{E}\left[\left(-\frac{1}{2} (\sigma_{t_i}^{\text{LN}})^2 \Delta t + \sigma_{t_i}^{\text{LN}} \Delta B_{t_i} \right)^2 \right]$$

where $\sigma_{t_i}^{\text{LN}}$ is the realized (log-normal) volatility between $[t_i, t_{i+1}]$, $\Delta B_{t_i} \equiv B_{t_{i+1}} - B_{t_i}$ and $T = n\Delta t$. This gives

$$\text{VS} \equiv \frac{1}{T} \sum_{i=0}^{n-1} \mathbb{E}\left[\left(\frac{1}{4} (\sigma_{t_i}^{\text{LN}})^4 (\Delta t)^2 + (\sigma_{t_i}^{\text{LN}})^2 \Delta t \right) \right]$$

By taking $\sigma_{t_i}^{\text{LN}} = \sigma^{\text{LN}}$ constant, we get

$$\sqrt{\text{VS}} \equiv (\sigma^{\text{LN}}) \left(1 + \frac{1}{4} (\sigma^{\text{LN}})^2 \Delta t \right)^{\frac{1}{2}}$$

If we impose a relative error of 10^{-3} between the continuous and the discrete version, we obtain $\Delta t = 8 \cdot 10^{-3} / (\sigma^{\text{LN}})^2$. For $\sigma^{\text{LN}} \sim 100\%$, we get $\Delta t \approx 3$ days.
□

At first sight, this payoff seems obviously model-dependent as $\mathbb{E}^{\mathbb{P}}[\langle \ln S \rangle_T]$ depends on $\mathbb{P} \in \mathcal{M}^c$. In fact, if we assume that T-Vanilla options are traded on the market, this option can be perfectly replicated and its (arbitrage-free) price is unique within the class $\mathcal{M}^c(\mu)$, as first observed in [67, 117]. This is established with the following pathwise equality (see Remark 3.2), obtained from Itô's lemma:

$$\frac{1}{T} \langle \ln S \rangle_T = -\frac{2}{T} \ln \frac{S_T}{S_0} + \frac{2}{T} \int_0^T \frac{dS_t}{S_t}, \quad \forall \mathbb{P} \in \mathcal{M}^c$$

This can be rewritten as: $\forall \, \mathbb{P} \in \mathcal{M}^c(\mu)$,

$$\frac{1}{T}\langle \ln S\rangle_T = -\mathbb{E}^\mu\left[\frac{2}{T}\ln\frac{S_T}{S_0}\right] + \left(-\frac{2}{T}\ln\frac{S_T}{S_0} + \mathbb{E}^\mu\left[\frac{2}{T}\ln\frac{S_T}{S_0}\right]\right) + \frac{2}{T}\int_0^T \frac{dS_t}{S_t}$$

This expression means that the variance swap payoff $\frac{1}{T}\langle \ln S\rangle_T$ can be replicated by statically hedging a T-Vanilla option with payoff $-2/T\ln\frac{S_T}{S_0}$ (and market price $-2/T\mathbb{E}^\mu[\ln\frac{S_T}{S_0}]$), by dynamically hedging the underlying with $H_t = 2/(TS_t)$ and holding a positive cash position $-2/T\mathbb{E}^\mu[\ln\frac{S_T}{S_0}]$.

REMARK 3.2   The stochastic integral $\int_0^T \frac{dS_t}{S_t}$ is typically defined $\mathbb{P}-$almost surely for some reference probability measure $\mathbb{P}$ under which the canonical process B is a semimartingale. In Section 4.1.1, we briefly review some attempts to extend this definition to any $\mathbb{P} \in \mathcal{M}^c$.   ▯

REMARK 3.3   In Chapter 4, we will consider the case when $\mathcal{M}^c$ is replaced by the space $\mathcal{M}^{\text{cadlag}}$ of càdlàg martingale probability measure. The variance swap is no more replicable (and therefore no more model-independent) within the space $\mathcal{M}^{\text{cadlag}}$. By taking the limit $n \to \infty$ of our discrete martingale Fréchet–Hoeffding solution (see Section 2.2.3), we will obtain the optimal martingale measures corresponding to the seller and buyer super-replication prices consistent with all Vanillas.   ▯

Our previous discussion can be generalized to the payoff $\frac{1}{2}\int_0^T \partial_s^2\omega(S_t)d\langle S\rangle_t$ that can be dynamically replicated by

$$\frac{1}{2}\int_0^T \partial_s^2\omega(S_t)d\langle S\rangle_t = \omega(S_T) - \omega(S_0) - \int_0^T \partial_s\omega(S_t)dS_t, \quad \forall \, \mathbb{P} \in \mathcal{M}^c$$

This implies that $\forall \, \mathbb{P} \in \mathcal{M}^c(\mu)$:

$$\mathbb{E}^\mathbb{P}\left[\frac{1}{2}\int_0^T \partial_s^2\omega(S_t)d\langle S\rangle_t\right] = \mathbb{E}^\mu[\omega(S_T) - \omega(S_0)]$$

$\partial_s^2\omega(s) \equiv 1_{s\in[A,B]}/s^2$ corresponds to a corridor variance swap.
In the next section, we explore other examples of nice martingales leading to model-independent payoffs (modulo the price of T-Vanillas).

3.2.2   Covariance options

We consider two liquid European options with payoffs F_1 and F_2 and maturity T, possibly depending on different assets. We denote $\mathbb{E}_t^\mathbb{P}[F_1]$ (resp. $\mathbb{E}_t^\mathbb{P}[F_2]$) the t-value of this option quoted on the market. The market uses a priori two (different) risk-neutral probability measures $\mathbb{P}^1$ and $\mathbb{P}^2$. We will assume that they coincide and belong to $\mathcal{M}^c$. $\mathbb{P}$ is not known, we have only a partial characterization through the values $\mathbb{E}_t^{\mathbb{P}}[F_1]$ and $\mathbb{E}_t^{\mathbb{P}}[F_2]$. We assume also that the payoff $F_1 F_2$ with maturity T can be bought at $t = 0$ with market prices $\mathbb{E}^{\mu}[F_1 F_2]$.

A covariance option pays at a maturity T the daily realized covariance between the prices $\mathbb{E}_t^{\mathbb{P}}[F_1]$ and $\mathbb{E}_t^{\mathbb{P}}[F_2]$:

$$\sum_{i=0}^{n-1} \left(\mathbb{E}_{t_{i+1}}^{\mathbb{P}}[F_1] - \mathbb{E}_{t_i}^{\mathbb{P}}[F_1] \right) \left(\mathbb{E}_{t_{i+1}}^{\mathbb{P}}[F_2] - \mathbb{E}_{t_i}^{\mathbb{P}}[F_2] \right)$$

In the limit $n \to \infty$, it converges to

$$\int_0^T d\langle \mathbb{E}^{\mathbb{P}}[F_1], \mathbb{E}^{\mathbb{P}}[F_2] \rangle_t$$

From Itô's lemma, we have for all $\mathbb{P} \in \mathcal{M}^c$:

$$\int_0^T d\langle \mathbb{E}^{\mathbb{P}}[F_1], \mathbb{E}^{\mathbb{P}}[F_2] \rangle_t = (F_1 F_2 - \mathbb{E}^{\mu}[F_1 F_2])$$

$$+ \left(\mathbb{E}^{\mu}[F_1 F_2] - \mathbb{E}_0^{\mathbb{P}}[F_1] \mathbb{E}_0^{\mathbb{P}}[F_2] \right)$$

$$- \int_0^T \mathbb{E}_t^{\mathbb{P}}[F_1] d\mathbb{E}_t^{\mathbb{P}}[F_2] - \int_0^T \mathbb{E}_t^{\mathbb{P}}[F_2] d\mathbb{E}_t^{\mathbb{P}}[F_1]$$

As observed in [76], this equality indicates that a covariance option can be replicated by doing a delta-hedging on $\mathbb{E}_t^{\mathbb{P}}[F_1]$ (resp. $\mathbb{E}_t^{\mathbb{P}}[F_2]$) with $H_t^1 \equiv -\mathbb{E}_t^{\mathbb{P}}[F_2]$ (resp. $H_t^2 \equiv -\mathbb{E}_t^{\mathbb{P}}[F_1]$) and statically holding the T-European payoff $F_1 F_2$ with market price $\mathbb{E}^{\mu}[F_1 F_2]$. The model-independent price of this option is therefore

$$\mathbb{E}^{\mathbb{P}}[\int_0^T d\langle \mathbb{E}^{\mathbb{P}}[F_1], \mathbb{E}^{\mathbb{P}}[F_2] \rangle_t] = \mathbb{E}^{\mu}[F_1 F_2] - \mathbb{E}_0^{\mathbb{P}}[F_1] \mathbb{E}_0^{\mathbb{P}}[F_2],$$

$$\forall\, \mathbb{P} \in \mathcal{M}^c \cap \{\mathbb{P} \,:\, \mathbb{E}^{\mathbb{P}}[F_1 F_2] = \mathbb{E}^{\mu}[F_1 F_2]\}$$

3.2.3 Lookback/Barrier options

We set $M_t \equiv \max_{s \in [0,t]} S_s$ the running maximum. We consider a smooth function $g(s, m) \in C^{2,1}(\mathbb{R}_+^2)$. By applying Itô's lemma,

$$dg(S_t, M_t) = \partial_s g(S_t, M_t) dS_t + \frac{1}{2} \partial_s^2 g(S_t, M_t) d\langle S \rangle_t + \partial_m g(S_t, M_t) dM_t$$

As M_t is an increasing process, it has a bounded variation and its quadratic variation vanishes. Furthermore, $dM_t \neq 0$ only when $S_t = M_t$ meaning that $\partial_m g(S_t, M_t) dM_t = \partial_m g(M_t, M_t) dM_t$. We deduce that

$$dg(S_t, M_t) = \partial_s g(S_t, M_t) dS_t + \frac{1}{2} \partial_s^2 g(S_t, M_t) d\langle S \rangle_t + \partial_m g(M_t, M_t) dM_t$$

$g(S_t, M_t)$ is then a local martingale if and only if its drift term vanishes:

$$\partial_s^2 g(s, m) = 0, \quad \forall\, s < m$$
$$\partial_m g(s, m)|_{s=m} = 0$$

from which we deduce the solutions:

THEOREM 3.2 Azéma–Yor martingale
Let S_t be a martingale, $a(\cdot) \in C^1(\mathbb{R})$ and $a^* \in \mathbb{R}$. Then

$$g_{AY}(S_t, M_t) = -(M_t - S_t)a(M_t) + \int_{a^*}^{M_t} a(u)du \tag{3.3}$$

is a martingale (called Azéma–Yor martingale [8, 9]).

We have only shown above that this is the only (local) martingale that can be written as a smooth function of S_t and its maximum M_t (see [119] where $a(\cdot)$ is only assumed to be a locally bounded function). By taking the expectation of both sides of (3.3), we get for all $\mathbb{P} \in \mathcal{M}^c$:

$$\mathbb{E}^{\mathbb{P}}[-(M_T - S_T)a(M_T) + \int_{a^*}^{M_T} a(u)du] = \mathbb{E}^{\mathbb{P}}[g_{AY}(S_t, M_t)] = g_{AY}(S_0, S_0)$$

For the last equality, we have used that $g_{AY}(S_t, M_t)$ is a martingale and $M_0 = S_0$. For $a^* = S_0$, this could be rewritten for all $\mathbb{P} \in \mathcal{M}^c(\mu)$ as

$$\mathbb{E}^{\mathbb{P}}[F(S_T, M_T)] = \mathbb{E}^{\mu}[\int_{S_0}^{S_T} a(y)dy] \tag{3.4}$$

where

$$F(S_T, M_T) \equiv a(M_T)(M_T - S_T) - \int_{S_T}^{M_T} a(y)dy$$

This relation indicates that the lookback option defined by the payoff F can be exactly replicated with Vanilla options with payoff $\int_{S_0}^{S_T} a(y)dy$ and a (zero cost) delta-hedging $H_t \equiv -a(M_t)$. Indeed, we have the pathwise equality: for all $\mathbb{P} \in \mathcal{M}^c(\mu)$,

$$F(S_T, M_T) = \int_{S_0}^{S_T} a(y)dy + \int_0^T H_t dS_t$$

and

$$\mathbb{E}^{\mathbb{P}}[F(S_T, M_T)] = \mathbb{E}^{\mu}[\int_{S_0}^{S_T} a(y)]$$

The market price $\mathbb{E}^\mu[\int_{S_0}^{S_T} a(y)]$ of $\int_{S_0}^{S_T} a(y)dy$ is inferred from T-Vanilla options (using formula (2.2)). An analogous result holds if we replace M_t by the running minimum $m_t \equiv \min_{s \in [0,t]} S_s$.

REMARK 3.4 If S_t was assumed to be a discontinuous martingale (in particular the running maximum is not continuous and $(M_t - S_t)dM_t \geq 0$), we would have for all $\mathbb{P} \in \mathcal{M}^{\text{cadlag}}$:

$$F(S_T, M_T) = \int_{S_0}^{S_T} a(y)dy + \int_0^T H_t dS_t + \int_0^T a'(M_t)(M_t - S_t)dM_t$$

$$\geq \int_{S_0}^{S_T} a(y)dy + \int_0^T H_t dS_t$$

with $a(\cdot)$ an increasing function. We have the reverse inequality when $a(\cdot)$ is decreasing. ▯

Example 3.1 Some identities
For particular choices of a, the identity (3.4) reduces to

$$\mathbb{E}^\mathbb{P}[(M_T - S_T)^2] = \mathbb{E}^\mu[(S_T - S_0)^2] \; : \; a(y) = 2y$$

$$\mathbb{E}^\mathbb{P}[1 - \frac{S_T}{M_T} + \ln \frac{S_T}{M_T}] = \mathbb{E}^\mu[\ln \frac{S_T}{S_0}] \; : \; a(y) = \frac{1}{y}$$

$$\mathbb{E}^\mathbb{P}[(B - M_T)1_{S_T \leq B \leq M_T}] = \mathbb{E}^\mu[\max(S_T, B) - \max(S_0, B)] \; : \; a(y) = 1_{y > B}$$

Does the left-hand side of the above relations look like Vanilla options to the reader?
Note that from the previous remark, $\mathbb{E}^\mathbb{P}[(M_T - S_T)^2] \geq \mathbb{E}^\mu[(S_T - S_0)^2]$ for all $\mathbb{P} \in \mathcal{M}^{\text{cadlag}}(\mu)$. ▯

Example 3.2 Portfolio's drawdown constraint
Let us consider a delta-hedge portfolio

$$d\Pi_t = H_t dS_t$$

For use below, we denote $\bar{\Pi}_t \equiv \max_{s < t} \Pi_s$. We want to find $H.$ such that the portfolio value satisfies the drawdown constraint:

$$\Pi_t \geq \alpha \bar{\Pi}_t \tag{3.5}$$

for some $\alpha \in (0, 1)$. We assume that $\Pi_t = \Pi(S_t, M_t)$. It turns out from Theorem 3.2 that Π_t is a delta-hedged portfolio if it is given by the Azéma–Yor martingale:

$$\Pi_t = -A'(M_t)(M_t - S_t) + A(M_t)$$

for some function A. If we take $A' > 0$ then

$$\bar{\Pi}_t = A(M_t)$$

The constraint (3.5) reads

$$-A'(m)(m - s) + A(m) \geq \alpha A(m), \quad \forall\, s < m$$

which is equivalent to

$$-A'(x)x + A(x) \geq \alpha A(x), \quad \forall\, x \in \mathbb{R}_+$$

A solution is given by $A(x) = x^{1-\alpha}$ for which

$$d\Pi_t = (1 - \alpha)M_t^{-\alpha}dS_t$$
$$= (1 - \alpha)(\Pi_t - \alpha\bar{\Pi}_t)\frac{dS_t}{S_t}$$

This can be extended to the case of a non-linear drawdown constraint as observed in [47]:

$$\Pi_t \geq \alpha(\bar{\Pi}_t) \tag{3.6}$$

with $\alpha(\cdot)$ a nonlinear function such that $\alpha(x) < x$. In the case $A' > 0$, this is equivalent to

$$-A'(x)x + A(x) \geq \alpha(A(x)), \quad \forall\, x \in \mathbb{R}_+$$

A solution is given by

$$\int_{\cdot}^{A(x)} \frac{dy}{y - \alpha(y)} = \ln x$$

Finally,

$$d\Pi_t = A'(A^{-1}(\bar{\Pi}_t))dS_t$$

$\square$

3.2.4   Options on spot/variance

A straightforward application of Itô's lemma shows that

$$\frac{l_t}{l_0} \equiv \left(\frac{S_t}{S_0}\right)^\beta e^{-\frac{1}{2}(\beta^2 - \beta)\langle \ln S \rangle_t}$$

is a $\mathbb{P}$-martingale for all $\mathbb{P} \in \mathcal{M}^c$. l_t has a nice financial interpretation as a Leveraged Exchange Trading Fund which has recently attracted the attention of investors. LETF tracks the value of an index, a basket of stocks, or another

ETF with the additional feature that it uses leverage. For a leverage ratio β, the dynamics of a LETF l_t is given by the following SDE:

$$\frac{dl_t}{l_t} = \beta \frac{dS_t}{S_t}$$

where S_t is the underlying ETF price. The identity

$$l_T = l_0 + \beta \int_0^T \frac{l_t}{S_t} dS_t$$

indicates that the payoff l_T can be perfectly replicated with a delta-hedging strategy $H_t = \beta l_t/S_t$.

By using the results of the previous section, we have that the Azéma–Yor payoff $a(M_T^l)(M_T^l - l_T) - \int_{l_T}^{M_T^l} a(y) dy$ is model-independent modulo the price of Vanillas on LETF, M^l being the maximum on l_t.

Suppose now that $f(t, x, y) \in C^{1,2,1}(\mathbb{R}_+^3)$ satisfies

$$\partial_t f(t, x, y) + \frac{1}{2} \partial_x^2 f(t, x, y) = 0$$
$$\partial_x f(t, 0, y) + \partial_y f(t, 0, y) = 0 \qquad (3.7)$$

Then a straightforward application of Itô's lemma shows that $f(\langle S \rangle_t, M_t - S_t, M_t)$ is a (local) martingale with $\langle S \rangle_t$ the quadratic variation of the spot. This specifies new examples of model-independent payoffs that can be perfectly replicated with a delta-hedging strategy:

$$f(\langle S \rangle_T, M_T - S_T, M_T) = f(\langle S \rangle_0, 0, M_0) - \int_0^T \partial_x f(\langle S \rangle_t, M_t - S_t, M_t) dS_t$$

We don't need to use static positions in T-Vanillas. As an example, take

$$f(\langle S \rangle_t, M_t - S_t, M_t) = (S_t - M_t)^2 - \langle S \rangle_t$$

3.2.5   Options on local time

Similarly, if f satisfies PDE (3.7), $f(\langle S \rangle_t, |S_t - S_0|, L_t)$ is a local martingale (and therefore specifies model-independent payoffs) where L_t is the local time of S at S_0[1]. Indeed,

$$df(\langle S \rangle_t, |S_t - S_0|, L_t) = \left(\partial_t f + \frac{1}{2} \partial_x^2 f \right) d\langle S \rangle_t + dL_t \left(\partial_x f(t, 0, y) + \partial_y f(t, 0, y) \right)$$
$$+ \partial_x f \left(1_{S_t > S_0} - 1_{S_t < S_0} \right) dS_t$$
$$= \partial_x f \left(1_{S_t > S_0} - 1_{S_t < S_0} \right) dS_t$$

---

[1] Formally, $dL_t = \delta(S_t - S_0) d\langle S \rangle_t$.

As an example, take

$$f(\langle S \rangle_t, |S_t - S_0|, L_t) = F(L_t) - |S_t - S_0| F'(L_t)$$

$F(L_T)$ can be interpreted in finance as a payoff depending at maturity T on the portfolio value of an at-the-money call option delta-hedged with its intrinsic value $(S_t - K)^+$. This is mathematically expressed by Tanaka's formula:

$$\frac{1}{2} L_T = (S_T - S_0)^+ - \int_0^T \mathbf{1}_{\{S_s > S_0\}} dS_s.$$

Note that L_T can also be interpreted as an infinitesimal corridor variance swap where we pay at the maturity T the sum of squared log-returns if S_t at time t belongs to a narrow corridor $[S_0 - \frac{\epsilon}{2}, S_0 + \frac{\epsilon}{2}]$:

$$L_T = \lim_{\epsilon \to 0} \frac{1}{\epsilon} \int_0^T \mathbf{1}_{[S_0 - \frac{\epsilon}{2}, S_0 + \frac{\epsilon}{2}]}(S_s) d\langle S \rangle_s$$

3.3 Timer options

We introduce a new class of model-independent options within the class $\mathcal{M}^c$. These options will pop up nicely when we will discuss the Skorokhod embedding problem in the next chapter.

3.3.1 Dirichlet options

A *timer-call* option pays to the holder a call option $(S-K)^+$ when the realized variance $V_t \equiv \langle \ln S \rangle_t$, defined as the quadratic variation of $\ln S$, reaches a budget $\bar{V}$. Here, no maturity is specified. These options were introduced by Société Génerale in 2007 [133].

A slight generalization consists in paying to the holder a payoff $g(S, V)$ when the variance V and the spot S reach the boundary of a domain $\mathcal{D}$ in $\mathbb{R}_+^2$. For a timer-call option, this corresponds to $g(S, V) = (S-K)^+$ and $\mathcal{D} = \mathbb{R}_+ \times [0, \bar{V}]$. We note below $\tau_{\mathcal{D}} \equiv \inf\{t > 0 : (S_t, V_t) \notin \mathcal{D}\}$ the first exit time from $\mathcal{D}$. As justified below, the (model-independent) fair value of this option, $u(S_0, V_0 \equiv 0)$, satisfies a Dirichlet PDE:

$$\partial_v u(s, v) + \frac{s^2}{2} \partial_s^2 u(s, v) = 0, \quad (s, v) \in \mathcal{D} \tag{3.8}$$

$$u(s, v) = g(s, v), \quad (s, v) \in \partial \mathcal{D}$$

The PDE (3.8) can be interpreted as an heat kernel equation (for a log-normal diffusion with unit volatility) with the variance playing the role of the time. $u(S_0, 0)$ is then the unique replication price.

Indeed, by applying Itô's lemma on $u(S_t, V_t) \in C^{2,1}$, we get, by using that u is a solution of PDE (3.8), for all $\mathbb{P} \in \mathcal{M}^c$:

$$g(S_{\tau_D}, V_{\tau_D}) = u(S_0, 0) + \int_0^{\tau_D} dV_t \left(\partial_v u(S_t, V_t) + \frac{S_t^2}{2} \partial_s^2 u(S_t, V_t) \right)$$
$$+ \int_0^{\tau_D} \partial_s u(S_t, V_t) dS_t$$
$$= u(S_0, 0) + \int_0^{\tau_D} \partial_s u(S_t, V_t) dS_t$$

This pathwise identity indicates that the payoff $g(S_{\tau_D}, V_{\tau_D})$ can be dynamically replicated with a delta-hedging $H_t = \partial_s u(S_t, V_t)$. The solution u does not depend on the volatility (see Equation (3.8)). As a striking consequence, the price and hedge of Dirichlet options are model-free within the class of arbitrage-free models with continuous paths (and with zero rates), which makes these products very convenient for hedging. From Dynkin's formula, we have the stochastic representation

$$u(s, v) = \mathbb{E}^{\mathbb{P}}[g(B_{\tau_D}^{\text{geo}}, \tau_D) | B_0^{\text{geo}} = s, V_0 = v] \tag{3.9}$$

with B^{geo} a geometric Brownian motion. Although the PDE (3.8) can be solved numerically with a Monte-Carlo simulation using the above-mentioned stochastic representation (3.9), we should observe that for particular payoffs and domains $\mathcal{D}$, u admits a closed-form formula - see below. Some of these closed-form formulas will be useful when we will exhibit solutions to SEP.

Examples

As already mentioned, a timer-call option corresponds to the domain $\mathcal{D} = \mathbb{R}_+ \times [0, \bar{V}]$. From (3.9), the solution is then given by the well-known Black–Scholes formula for a call option with variance $\bar{V}$:

$$u(S_0, V_0 = 0) = C_{\text{BS}}(\sigma^2 T = \bar{V}, K)$$

As a new example of timer options, we consider the payoff $\exp(\lambda S)$ that we deliver when the quadratic variation of the spot reaches the target $V \geq \bar{V}$. Here for convenience, we take $V_t \equiv \langle S \rangle_t$ and the PDE (3.8) is replaced by

$$\partial_v u(s, v) + \frac{1}{2} \partial_s^2 u(s, v) = 0$$
$$u(s, \bar{V}) = e^{\lambda s}$$

The solution is $u(s, v) = e^{\lambda s - \frac{\lambda^2}{2}(v - \bar{V})}$ for which we deduce the nice identity

$$\mathbb{E}^{\mathbb{P}}[e^{\lambda(S_{\tau_{\bar{V}}} - S_0)}] = e^{\frac{\lambda^2}{2} \bar{V}}, \quad \forall \mathbb{P} \in \mathcal{M}^c$$

with $\tau_{\bar{V}} \equiv \inf\{t > 0 : V_t \geq \bar{V}\}$. This identity indicates that $S_{\tau_{\bar{V}}}$ has the same characteristic function as a Brownian motion $B_{\bar{V}}$. This result, known as Dambis-Dubins-Schwarz theorem (see e.g. [130] for a proof), states precisely that a continuous local martingale evaluated at its quadratic variation is nothing else than a Brownian motion:

THEOREM 3.3 Dambis–Dubins–Schwarz (in short DDS)
Let S_t be a continuous $\mathcal{F}_t$-local martingale, satisfying $\lim_{t\to\infty}\langle S\rangle_t = \infty$ $\mathbb{P}$-a.s. Define for each $0 \leq \bar{V} \leq \infty$ the stopping time $\tau_{\bar{V}} = \inf\{t \geq 0 : \langle S\rangle_t \geq \bar{V}\}$. Then the time-changed process $B_{\bar{V}} = S_{\tau_{\bar{V}}}$ is a $\mathcal{G}_{\bar{V}}$-Brownian motion where $\mathcal{G}_{\bar{V}} = \mathcal{F}_{\tau_{\bar{V}}}$. In particular, we have $\mathbb{P}$-a.s., $S_t = B_{\langle S\rangle_t}$.

3.3.2 Neumann options

Let us first consider an option which pays a payoff $g(S_t, M_t)$ when (S_t, M_t) reaches a boundary $\partial\mathcal{D}$. The fair value of this option, $u(S_0, S_0)$, is solution of the Dirichlet–Neumann problem:

$$\partial_s^2 u(s, m) = 0, \quad \{s < m\} \cap \mathcal{D} \tag{3.10}$$

$$\partial_m u(s, m)|_{s=m} = 0 \tag{3.11}$$

$$u(s, m) = g(s, m), \quad (s, m) \in \partial\mathcal{D} \tag{3.12}$$

which is obviously model-independent. $u(S_0, M_0 \equiv S_0)$ can be interpreted as a replication price. Indeed, by applying Itô's lemma on $u \in C(\overline{\mathcal{D}}) \cap C^2(\mathcal{D})$, we have

$$g(S_{\tau_\mathcal{D}}, M_{\tau_\mathcal{D}}) = u(S_0, S_0) + \int_0^{\tau_\mathcal{D}} \left(\frac{1}{2}\partial_s^2 u(S_t, M_t)d\langle S\rangle_t + \partial_M u(S_t, M_t)dM_t \right)$$

$$+ \int_0^{\tau_\mathcal{D}} \partial_s u(S_t, M_t)dS_t$$

$$= u(S_0, S_0) + \int_0^{\tau_\mathcal{D}} \partial_s u(S_t, M_t)dS_t$$

indicating that the payoff $g(S_{\tau_\mathcal{D}}, M_{\tau_\mathcal{D}})$ can be dynamically replicated with a delta $H_t = \partial_s u(S_t, M_t)$. We have used from the Neumann condition (3.11) that

$$\partial_M u(S_t, M_t)dM_t = \partial_M u(M_t, M_t)dM_t = 0$$

3.3.2.1 Example: Barrier options

As an example, let us take $\mathcal{D} = (\underline{s}, \bar{s}) \times \mathbb{R}_+$. Here equation (3.12) reads

$$u(\underline{s}, m) = g(\underline{s}, m) \tag{3.13}$$

$$u(\bar{s}, \bar{s}) = g(\bar{s}, \bar{s}) \tag{3.14}$$

as $m = \overline{s}$ when the spot reaches the upper barrier $s = \overline{s}$. Then, a straightforward computation gives

$$u(s,m) = \left(-\int_{\overline{s}}^{m} \frac{\partial_m g(\underline{s}, x)}{x - \underline{s}} dx + \frac{g(\overline{s}, \overline{s}) - g(\underline{s}, \overline{s})}{\overline{s} - \underline{s}} \right) (s - \underline{s}) + g(\underline{s}, m)$$

and therefore

$$\mathbb{E}[g(S_{\tau_D})] = \left(-\int_{\overline{s}}^{S_0} \frac{\partial_m g(\underline{s}, x)}{x - \underline{s}} dx + \frac{g(\overline{s}, \overline{s}) - g(\underline{s}, \overline{s})}{\overline{s} - \underline{s}} \right) (S_0 - \underline{s}) + g(\underline{s}, S_0)$$

3.3.2.2 Example: Azéma–Yor domain

We define $\tau_{\text{AY}} \equiv \inf\{t > 0 \,:\, S_t \leq \psi(M_t)\}$ where ψ is a monotone function with $\psi(x) < x$. The solution of the Neumann option $\mathbb{E}[\lambda(S_{\tau_{\text{AY}}})]$ can be solved analytically:

LEMMA 3.1

$$\mathbb{E}[\lambda(S_{\tau_{\text{AY}}})] = \left(\int_{S_0}^{\infty} \frac{\lambda'(\psi(x))}{x - \psi(x)} d\psi(x) e^{\int_x^{S_0} \frac{d\psi(y)}{y - \psi(y)}} \right) (S_0 - \psi(S_0)) + \lambda(\psi(S_0))$$

PROOF Set $u(s,m) \equiv \mathbb{E}[\lambda(S_{\tau_{\text{AY}}})|S_0 = s, M_0 = m]$. u satisfies PDE:

$$\begin{cases} \partial_s^2 u(s,m) = 0, & s \leq m \\ \partial_m u(m,m) = 0 \\ u(\psi(m),m) = \lambda(\psi(m)) \end{cases}$$

for which the solution is

$$u(s,m) = \left(C e^{\int_{S_0}^{m} \frac{\psi'(y)}{y - \psi(y)} dy} - \int_{S_0}^{m} \frac{\lambda'(\psi(x))}{x - \psi(x)} d\psi(x) e^{\int_x^{m} \frac{d\psi(y)}{y - \psi(y)}} \right) (s - \psi(m))$$
$$+ \lambda(\psi(m))$$

with C an integration constant. u should stay finite when m goes to infinity. This fixes C and we have:

$$u(s,m) = \left(\int_{m}^{\infty} \frac{\lambda'(\psi(x))}{x - \psi(x)} d\psi(x) e^{\int_x^{m} \frac{d\psi(y)}{y - \psi(y)}} \right) (s - \psi(m)) + \lambda(\psi(m))$$

Finally, by taking $s = m = S_0$, we get our result. □

3.3.2.3 Example: Range timer options

Define the range process as $R_t \equiv M_t - m_t$ with $m_t = \min_{s \leq t} S_s$ and the stopping time $\tau_L \equiv \inf\{t \geq 0 \,:\, R_t \geq L\}$ where $L > 0$ is a given threshold. A range timer option pays to the holder a payoff $g(S)$ on the spot when the range reaches the level L. Proceeding as above, we obtain that the unique replication value $u(S_0, S_0, S_0)$ is solution (with the requirement that $u \in C(\overline{\mathcal{D}}) \cap C^2(\mathcal{D})$) to the Dirichlet–Neumann problem:

$$\partial_s^2 u(s, m, M) = 0, \quad m < s < M$$
$$\partial_M u(s, m, M)|_{s=M} = 0$$
$$\partial_m u(s, m, M)|_{s=m} = 0$$
$$u(m, m, L + m) = g(m)$$
$$u(M, M - L, M) = g(M)$$

which is obviously model-independent. Here m (resp. M) denotes the variable associated to the minimum (resp. maximum). By integrating these simple equations (see [87] for details), we obtain that the solution is

$$u(S_0, S_0, S_0) = L^{-2} \left(\int_{S_0-L}^{S_0} (S_0 - x) g(x) dx + \int_{S_0}^{S_0+L} (x - S_0) g(x) dx \right)$$

3.3.3   Some generalizations

The above section can be easily generalized by considering a payoff depending on the spot, its minimum, its maximum and its realized variance. Its fair value $u(S_0, S_0, S_0, V_0 \equiv 0)$ is solution of the model-independent pricing PDE:

$$\partial_v u(s, m, M, v) + \frac{s^2}{2} \partial_s^2 u(s, m, M, v) = 0, m < s < M$$
$$\partial_M u|_{s=M} = 0, \quad \partial_m u|_{s=m} = 0, \quad u_{\partial \mathcal{D}} = g$$

Below, we consider a timer-like option depending on the local time.

3.3.3.1   Example: Vallois domain

Set $\tau_{\text{Vallois}} \equiv \inf\{t > 0 : S_t \notin (\phi_-(L_t), \phi_+(L_t))\}$, ϕ_+ (ϕ_-) increasing (decreasing) and $\phi_-(0) = \phi_+(0) = S_0$. We recall that L_t denotes the local time at S_0. The solution of the timer-like option $\mathbb{E}[\lambda(S_{\tau_{\text{Vallois}}})]$ can be solved analytically:

LEMMA 3.2

Let $\phi_+ : \mathbb{R}_+ \to [S_0, \infty)$ (resp. $\phi_- : \mathbb{R}_+ \to (0, S_0]$) be an increasing (resp. decreasing) function such that $\gamma(0+) = 0$ and $\gamma(+\infty) = +\infty$, where

$$\gamma(l) \equiv \frac{1}{2} \int_0^l \left(\frac{1}{\phi_+(m) - S_0} + \frac{1}{S_0 - \phi_-(m)} \right) dm \text{ for all } l > 0.$$

Then for $\lambda \in L^\infty$,

$$\mathbb{E}[\lambda(S_{\tau_{\text{Vallois}}})] = \frac{1}{2} \int_0^{+\infty} \left(\frac{\lambda(\phi_+(m))}{\phi_+(m) - S_0} + \frac{\lambda(\phi_-(m))}{S_0 - \phi_-(m)} \right) e^{-\gamma(m)} dm$$

PROOF   We denote

$$c(l) = \frac{1}{2} \int_l^{+\infty} \left(\frac{\lambda(\phi_+(m))}{\phi_+(m) - S_0} - \frac{\lambda(\phi_-(m))}{\phi_-(m) - S_0} \right) e^{\gamma(l) - \gamma(m)} dm \quad \text{for all } l \geq 0.$$

Let $(M_t)_{t \geq 0}$ be the process given by

$$M_t = \frac{\lambda(\phi_+(L_t)) - c(L_t)}{\phi_+(L_t) - S_0} (S_t - S_0)^+ - \frac{\lambda(\phi_-(L_t)) - c(L_t)}{\phi_-(L_t) - S_0} (S_t - S_0)^- + c(L_t).$$

By applying Tanaka's formula, we deduce that

$$M_t = M_0$$
$$+ \int_0^t \left(\frac{\lambda(\phi_+(L_s)) - c(L_s)}{\phi_+(L_s) - S_0} \mathbf{1}_{\{S_s > S_0\}} - \frac{\lambda(\phi_-(L_s)) - c(L_s)}{\phi_-(L_s) - S_0} \mathbf{1}_{\{S_s \leq S_0\}} \right) dS_s$$

Hence, the process M is a local martingale. Further, the stopped process[2] $M^{\tau_{\text{Vallois}}} \equiv (M_{t \wedge \tau_{\text{Vallois}}})_{t \geq 0}$ is bounded by $5\|\lambda\|_\infty$ since $\|c\|_\infty \leq \|\lambda\|_\infty$ and $(S_{\tau \wedge t} - S_0)^\pm \leq |\phi_\pm(L_{\tau \wedge t}) - S_0|$. It follows that

$$\mathbb{E}_{s,l}[M_{\tau_{\text{Vallois}}}] = \frac{\lambda(\phi_+(l)) - c(l)}{\phi_+(l) - S_0} (s - S_0)^+ - \frac{\lambda(\phi_-(l)) - c(l)}{\phi_-(l) - S_0} (s - S_0)^- + c(l).$$

To conclude, it remains to see that $M_{\tau_{\text{Vallois}}} = \lambda(S_{\tau_{\text{Vallois}}})$ by definition of τ_{Vallois}. ☐

3.3.4   Model-dependence

3.3.4.1   With rates

Up to this point, we have assumed zero rates. This can be easily relaxed by replacing the spot S_t by its forward $f_t \equiv S_t e^{-rt}$ which is driftless, M_t by the running maximum of the forward, $\max_{s \leq t} f_s$, and V_t by $\langle \ln f \rangle_t$. Otherwise, the model-dependence of timer options breaks down. For example, in the case of a time-homogeneous local volatility model, for which $dS_t = S_t \sigma(S_t) dB_t + rS_t dt$, and a constant interest rate r, we have that the (no-robust) replication price $u(S_0, 0)$ of a timer call option is solution of

$$\sigma(s)^2 \left(\partial_v u(s, v) + \frac{s^2}{2} \partial_s^2 u(s, v) \right) + rs \partial_s u(s, v) = ru(s, v), \quad u(s, \bar{V}) = g(s)$$

From this PDE, we see the dependence of u on the local volatility function $\sigma(\cdot)$ (when $r \neq 0$).

---

[2]$a \wedge b \equiv \min(a, b)$.

3.3.4.2 With finite horizon

In practice, a maturity T is specified for a timer-like option - the holder does not want to wait possibly forever. For example, for a timer, the payoff reads $g(S_{\tau \wedge T}, V_{\tau \wedge T})$. The model-dependence breaks down and can be characterized by the following proposition:

PROPOSITION 3.1
We set $u_{\text{timer}}(s,v) \equiv \mathbb{E}[g(S_\tau, V_\tau)|S_0 = s, V_0 = v]$ with $\tau = \inf\{t > 0 : (S_t, V_t) \notin \mathcal{D}\}$ and $\mathcal{D} = [0, \bar{V}] \times \mathbb{R}_+$. Then, for all $\mathbb{P} \in \mathcal{M}^c$,

$$\mathbb{E}^{\mathbb{P}}[g(S_{\tau \wedge T}, V_{\tau \wedge T})] = u_{\text{timer}}(S_0, 0) + \mathbb{E}^{\mathbb{P}}[(g - u_{\text{timer}})(S_T, V_T)]$$

PROOF   We have

$$\begin{aligned}
\mathbb{E}[g(S_{\tau \wedge T}, V_{\tau \wedge T}) - g(S_\tau, V_\tau)] &= \mathbb{E}[(g(S_T, V_T) - g(S_\tau, V_\tau))\,1_{\tau \geq T}] \\
&= \mathbb{E}[(g(S_T, V_T) - \mathbb{E}_T[g(S_\tau, V_\tau)])\,1_{\tau \geq T}] \\
&\overset{\text{Strong Markov}}{=} \mathbb{E}[(g(S_T, V_T) - u_{\text{timer}}(S_T, V_T))\,1_{\tau \geq T}] \\
&= \mathbb{E}[(g - u_{\text{timer}})(S_T, V_T)]
\end{aligned}$$

as $g(S_T, V_T) = u_{\text{timer}}(S_T, V_T)$ for $\tau < T$.   ⬜

The model-dependence of the T-timer options is synthesized by the payoff $(g - u_{\text{timer}})(S_T, V_T)$. The above proposition still holds if the exit time satisfies the condition: $T > \tau \implies (S_T, V_T) \notin \mathcal{D}$.

Example 3.3 Timer call
For a T-timer call option with $g(s) = (s - K)^+$, we obtain that for all $\mathbb{P} \in \mathcal{M}^c(\mu)$,

$$\begin{aligned}
\mathbb{E}^{\mathbb{P}}[(S_{\tau_{\bar{V}} \wedge T} - K)^+] = \ &C_{BS}(\sigma^2 T \equiv \bar{V}, K) \\
&- \mathbb{E}[C_{BS}(\sigma^2 T \equiv (\bar{V} - V_T)^+, K)] + \mathbb{E}^{\mu}[(S_T - K)^+]
\end{aligned}$$

A T-timer call option boils down to the (model-dependent) pricing and hedging of an option on the spot and the variance with payoff $C_{BS}(\sigma^2 T \equiv (\bar{V} - V_T)^+, K)$ and a static hedge in a T-call option with strike K.
Similarly, for a T-timer p-payoff with $g = S^p$, we have for all $\mathbb{P} \in \mathcal{M}^c(\mu)$:

$$\mathbb{E}[S^p_{\tau_{\bar{V}} \wedge T}] = S^p_0 e^{\frac{p(p-1)\bar{V}}{2}} - \mathbb{E}[S^p_T e^{\frac{p(p-1)}{2}(\bar{V} - V_T)^+}] + \mathbb{E}^{\mu}[S^p_T]$$

⬜

3.4 Ocone's martingales

Under the assumption of zero rates, we have characterized model-independent payoffs, invariant by time-change, that can be perfectly replicated by dynamic hedging, possibly with a static position in T-Vanillas for all $\mathbb{P} \in \mathcal{M}^c(\mu)$. In the rest of this chapter, we will show that the class of model-independent payoffs can be enlarged by restricting the class $\mathcal{M}^c(\mu)$ to the class of Ocone's martingales with fixed T-marginal μ: $\mathcal{M}^c(0, \mu) \subset \mathcal{M}^c(\mu)$.

Class $\mathcal{M}^c(0)$: For any $\mathcal{F}$-adapted process σ_t and satisfying $\int_0^T \sigma_s^2 ds < \infty$, $\mathbb{P}^0$-a.s. and $\lim_{T \to \infty} \int_0^T \sigma_s^2 ds = \infty$, we define the martingale measure on $(\Omega, \mathcal{F})$,

$$\mathbb{P}^\sigma \equiv \mathbb{P}^0 \circ (S^\sigma)^{-1}$$

where

$$dS_t^\sigma = \sigma_t dB_t, \quad t \in [0, T], \mathbb{P}^0 - \text{a.s.}$$

with σ and B *independent*. We denote by $\mathcal{M}^c(0) \subset \mathcal{M}^c$ the collection of all such probability measures on $(\Omega, \mathcal{F})$ and $\mathcal{M}^c(0, \mu) \equiv \mathcal{M}^c(0) \cap \mathcal{M}^c(\mu)$. The notation $\mathcal{M}^c(0)$ indicates the uncorrelation between σ. and B..

Note that in order to restrain S_t^σ to be positive, we could have defined the class $\mathcal{M}^c(0)$ as the processes for which

$$dS_t^\sigma = \sigma_t S_t^\sigma dB_t$$

with σ and B *independent*. We skip this case for the sake of simplicity (but all our results can be easily extended to this setting). S_t^σ is called an Ocone martingale.

An alternative definition is

DEFINITION 3.1 Ocone's martingale *Let S_t be a continuous $\mathcal{F}_t$-local martingale, satisfying $\lim_{t \to \infty} \langle S \rangle_t = \infty$ $\mathbb{P}$-a.s. Consider its DDS representation: $S_t = B_{\langle S \rangle_t}$. The process S_t is called an Ocone martingale if B_t and $\langle S \rangle_t$ are independent.*

As previously, for the ease of notation, we will delete the superscript σ on S^σ below.

We have the following lemma:

LEMMA 3.3
For all $\mathbb{P} \in \mathcal{M}^c(0)$, S_t is conditionally symmetric.

DEFINITION 3.2 Conditionally symmetric   *Let S_t be an adapted process. S_t is conditionally symmetric if for any $\tau \in [0, T]$ and any non-negative Borel function f*

$$\mathbb{E}[f(S_T - S_\tau)|\mathcal{F}_\tau] = \mathbb{E}[f(S_\tau - S_T)|\mathcal{F}_\tau]$$

PROOF of Lemma 3.3   We have

$$\mathbb{E}^{\mathbb{P}}[f(S_T - S_\tau)] = \mathbb{E}^{\mathbb{P}}[f\left(\int_\tau^T \sigma_s dB_s\right)] = \mathbb{E}^{\mathbb{P}}[\mathbb{E}^{\mathbb{P}}[f\left(\int_\tau^T \sigma_s dB_s\right)|\mathcal{F}^\sigma]]$$

where $\mathcal{F}^\sigma$ is the filtration generated by the volatility σ. From the independence of σ and B and the fact that conditional to $\mathcal{F}^\sigma$, $-\int_\tau^T \sigma_s dB_s$ has the same law as $\int_\tau^T \sigma_s dB_s$ (i.e., Gaussian distribution), we deduce our result.   ☐

This result indicates that the Bachelier implied volatility[3] σ_B at a time τ for a maturity $T > \tau$ is symmetric as a function of the strike: $\sigma_B(T, K) = \sigma_B(T, -K)$. This property can therefore be checked or invalidated by observing market implied volatilities. The author has never observed a symmetric implied volatility in equity markets (which are skewed) except in the foreign-exchange markets.

A classification of martingales: Complete/incomplete models

In Theorem 3.1, we have characterized the extremal points of $\mathcal{M}^c$ and see that they correspond to complete models. An Ocone martingale, corresponding to an uncorrelated stochastic volatility model, is therefore an incomplete model, except in the trivial case where it is a Bachelier model (the stochastic volatility is here a deterministic function). A Bachelier model defines an example of pure martingales.

DEFINITION 3.3 Pure martingale   *Let S_t be a continuous $\mathcal{F}_t$-local martingale, satisfying $\lim_{t\to\infty}\langle S \rangle_t = \infty$ $\mathbb{P}$-a.s. Consider its DDS representation: $S_t = B_{\langle S \rangle_t}$. If $\langle S \rangle_t$ is measurable for the filtration $\mathcal{F}^B_\infty$ generated by B for every $t \geq 0$, then S_t is called a pure martingale.*

We denote $\mathcal{M}^{\text{pure}}$ the class of probability measures $\mathbb{P}$ such that S_t is a $\mathbb{P}$-pure martingale. The local volatility model is an example of pure martingale. The following proposition (illustrated by Figure 3.1) shows that pure martingales correspond to complete models.

---

[3]This is defined as the volatility σ such that $C(T, K) = C_B(\sigma^2 T, K)$ with $C_B(\sigma^2 T, K) = \frac{1}{2}(S_0 - K)\text{erfc}\left(\frac{K - S_0}{\sqrt{2}\sigma\sqrt{T}}\right) + \frac{\sigma\sqrt{T}}{\sqrt{2\pi}}e^{-\frac{(K-S_0)^2}{2\sigma^2 T}}$. $C_B(\sigma^2 T, K)$ is obtained by replacing the Black–Scholes log-normal process by a normal process with a constant volatility σ: $dS_t = \sigma dB_t$.

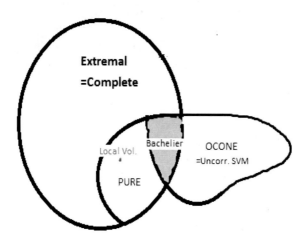

Figure 3.1: A sketch of a martingale classification drawn from [143].

PROPOSITION 3.2 [143]

(i) $\mathcal{M}^{\mathrm{pure}} \subset \mathrm{Ext}(\mathcal{M}^c)$.
(ii) $\mathcal{M}^{\mathrm{pure}} \cap \mathcal{M}^c(0)$ *is the set of Gaussian martingales (i.e., Bachelier model* $dS_t = \sigma(t)dB_t$ *for a deterministic volatility* $\sigma(t)$*).*

PROOF First, we show that the law $\mathbb{P}^B$ of the Brownian motion B is extremal. Assume that

$$\mathbb{P}^B = \theta\mathbb{P} + (1-\theta)\mathbb{Q}, \quad \theta \in (0,1), \quad \mathbb{P}, \mathbb{Q} \in \mathcal{M}^c$$

Then[4] $\mathbb{P} \leq \mathbb{P}^B$. We have $\langle B \rangle_t = t$ $\mathbb{P}^B$-a.s. This implies that $\langle B \rangle_t = t$ $\mathbb{P}$-a.s. From Lévy's characterisation of Brownian motion (see e.g. [130]), this implies that B is a Brownian motion and $\mathbb{P} = \mathbb{P}^B$.
(i) Let S be a pure martingale: $S_t = B_{\langle S \rangle_t}$. As S is generated by B, a functional $F_T(S_u, u \geq 0)$ can be written from the extremality of the law of B as

$$F = C + \int_0^\infty H_t(B)dB_t = C + \int_0^\infty H_{\langle S \rangle_t}(B)dS_t$$

This implies that the distribution of S is extremal from Theorem 3.1.
(ii) $\langle S \rangle_t$ must be measurable with respect to B and independent of B. It is therefore deterministic and we get a Gaussian martingale. $\qquad\square$

---

[4]$\mathbb{P} \leq \mathbb{Q}$ if $\mathbb{Q}(A) = 0$ for $A \in \mathcal{F}$ implies $\mathbb{P}(A) = 0$.

3.4.1 Lookback/Barrier options

PROPOSITION 3.3 [41, 46]
Let G be an arbitrary function.
(i) *Pathwise equality: For all* $\mathbb{P} \in \mathcal{M}^c(0)$

$$\Gamma(S_T) + 1_{\tau_B \leq T} \left(G(S_T) 1_{S_T \leq B} - G(2B - S_T) 1_{S_T > B} \right) = G(S_T) 1_{M_T \geq B}$$

with $\Gamma(S_T) \equiv (G(S_T) + G(2B - S_T)) 1_{S_T > B}$ *and* $\tau_B \equiv \inf\{t > 0 : S_t \geq B\}$.
(ii): *For all* $\mathbb{P} \in \mathcal{M}^c(0, \mu)$, *we have*

$$\mathbb{E}^{\mathbb{P}}[G(S_T) 1_{M_T \geq B}] = \mathbb{E}^{\mu}[\Gamma(S_T)]$$

PROOF
(i): The pathwise equality can be checked on a case-by-case basis: (a) If $\tau_B > T$, we get $0 = 0$. (b) If $\tau_B \leq T, S_T \leq B$, we get $G(S_T) = G(S_T)$. (c) If $\tau_B \leq T, S_T > B$, we have $G(S_T) + G(2B - S_T) - G(2B - S_T) = G(S_T)$.
(ii): Taking the expectation on both sides of this pathwise equality, we get

$$\mathbb{E}^{\mu}[\Gamma(S_T)] + \mathbb{E}^{\mathbb{P}}[1_{\tau_B \leq T} \left(G(S_T) 1_{S_T \leq B} - G(2B - S_T) 1_{S_T > B} \right)]$$
$$= \mathbb{E}^{\mathbb{P}}[G(S_T) 1_{M_T \geq B}]$$

From the strong Markov property, we have

$$\mathbb{E}^{\mathbb{P}}[1_{\tau_B \leq T} \left(G(S_T) 1_{S_T \leq B} - G(2B - S_T) 1_{S_T > B} \right)] =$$
$$\mathbb{E}^{\mathbb{P}}[1_{\tau_B \leq T} \mathbb{E}[(G(S_T) 1_{S_T \leq B} - G(2B - S_T) 1_{S_T > B}) | S_{\tau_B} = B]]$$

The right-hand side cancels from the conditional symmetry property (see Lemma 3.3) and we conclude. □

The pathwise equality has a nice hedging interpretation: the payoff $G(S_T) 1_{M_T \geq B}$ can be perfectly replicated within the class $\mathcal{M}^c(0)$ by (1) a static hedge with a T-Vanilla with payoff $\Gamma(S_T)$ and (2) selling a payoff $G(2B - S_T) 1_{S_T > B}$ and buying a payoff $G(S_T) 1_{S_T \leq B}$ at τ_B. From the conditional symmetry property, the hedge (2) is done at zero cost. The model-independent price of $G(S_T) 1_{M_T \geq B}$ corresponds therefore to the price of $\Gamma(S_T)$, i.e., $\mathbb{E}^{\mu}[\Gamma(S_T)]$.

Example 3.4 One-touch digital
In the particular case, $G(S_T) = 1$, this reduces to

$$\mathbb{E}^{\mathbb{P}}[1_{M_T \geq B}] = 2\mathbb{E}^{\mu}[1_{S_T > B}]$$

By integration, $\mathbb{E}^{\mathbb{P}}[(M_T - B)^+] = 2\mathbb{E}^{\mu}[(S_T - B)^+]$. □

3.4.2 Options on variance

PROPOSITION 3.4 [44]

(i) *Pathwise equality: For all* $\mathbb{P} \in \mathcal{M}^c(0)$, *we have*

$$e^{\frac{p^2}{2}\langle S \rangle_T} = \mathbb{E}^{\mathbb{P}}[e^{p(S_T - S_0)}] + \int_0^T e^{\frac{p^2}{2}\langle S \rangle_t - pS_t} d\mathbb{E}_t^{\mathbb{P}}[e^{pS_T}]$$

$$- p \int_0^T e^{\frac{p^2}{2}\langle S \rangle_t - pS_t} \mathbb{E}_t^{\mathbb{P}}[e^{pS_T}] dS_t \qquad (3.15)$$

(ii): *For all* $\mathbb{P} \in \mathcal{M}^c(0, \mu)$, *we have*

$$\mathbb{E}^{\mathbb{P}}[e^{\frac{p^2}{2}\langle S \rangle_T}] = \mathbb{E}^{\mu}[e^{p(S_T - S_0)}]$$

This means that the payoff $e^{\frac{p^2}{2}\langle S \rangle_T}$ can be perfectly replicated with a unique price $\mathbb{E}^{\mu}[e^{p(S_T - S_0)}]$ using a dynamic hedging in the spot S_t and in a Vanilla option $\mathbb{E}_t^{\mathbb{P}}[e^{pS_T}]$.

PROOF We have the identity $\forall\, \mathbb{P} \in \mathcal{M}^c(0)$:

$$\mathbb{E}_t^{\mathbb{P}}[e^{p(S_T - S_t)}] = \mathbb{E}_t^{\mathbb{P}}[e^{p(\int_t^T \sigma_s dB_s)}]$$

$$= \mathbb{E}_t^{\mathbb{P}}[\mathbb{E}[e^{p(\int_t^T \sigma_s dB_s)}|\mathcal{F}^{\sigma}]] = \mathbb{E}_t^{\mathbb{P}}[\mathbb{E}[e^{\frac{p^2}{2}(\int_t^T \sigma_s^2 ds)}|\mathcal{F}^{\sigma}]]$$

$$= \mathbb{E}_t^{\mathbb{P}}[e^{\frac{p^2}{2}(\langle S \rangle_T - \langle S \rangle_t)}]$$

where we have used that conditional on the volatility, $\int_t^T \sigma_s dB_s$ is a Gaussian random variable with variance $\int_t^T \sigma_s^2 ds$. For $t = 0$, this reproduces (ii). By applying Itô's lemma on $\mathbb{E}_t^{\mathbb{P}}[e^{\frac{p^2}{2}\langle S \rangle_T}] = e^{\frac{p^2}{2}\langle S \rangle_t - pS_t} \mathbb{E}_t^{\mathbb{P}}[e^{pS_T}]$, this implies Equation (3.15). $\qquad \square$

REMARK 3.5 Note that the replication strategy (3.15) is valid if and only if

$$\int_0^T e^{\frac{p^2}{2}\langle S \rangle_t - pS_t} d\langle \mathbb{E}^{\mathbb{P}}[e^{pS_T}], S \rangle_t = 0$$

This means that $S.$ and $\mathbb{E}^{\mathbb{P}}.[e^{pS_T}]$ should be uncorrelated. This can be easily derived by integrating by parts the term $\int_0^T e^{\frac{p^2}{2}\langle S \rangle_t - pS_t} d\mathbb{E}_t^{\mathbb{P}}[e^{pS_T}]$. Of course, this condition is valid if and only if $\mathbb{P} \in \mathcal{M}^c(0)$. $\qquad \square$

3.4.3   Options on local time

Below, we will show that by restricting the set $\mathcal{M}^c(\mu)$ to the set $\mathcal{M}^c(0, \mu)$, a call option $(L_T - K)^+$ on the local time with strike K can be dynamically replicated with a dynamic hedge at t on a T-call option $c_t(S_0 + (K - L_t)^+)$ with strike $S_0 + (K - L_t)^+$ and a T-put option $p_t(S_0 - (K - L_t)^+)$ with strike $S_0 - (K - L_t)^+$ - see additional results in [40].

PROPOSITION 3.5 [40]

(i) *Pathwise equality: For all* $\mathbb{P} \in \mathcal{M}^c(0)$,

$$(L_T - K)^+ = 2\mathbb{E}^{\mathbb{P}}[(S_T - S_0 - K)^+]$$

$$+2 \int_0^T \left(1_{S_t > S_0}(dp_t)(S_0 - (K - L_t)^+) + 1_{S_t \leq S_0}(dc_t)(S_0 + (K - L_t)^+) \right)$$

(ii) *For all* $\mathbb{P} \in \mathcal{M}^c(0, \mu)$, *we have*

$$\mathbb{E}^{\mathbb{P}}[(L_T - K)^+] = 2\mathbb{E}^{\mu}[(S_T - S_0 - K)^+] \qquad (3.16)$$

More generally for a payoff F, *we have*

$$\mathbb{E}^{\mathbb{P}}[F(L_T)] = F(0) + 2F'(0)\mathbb{E}^{\mu}[(S_T - S_0)^+] + 2 \int_0^\infty F''(dK)\mathbb{E}^{\mu}[(S_T - S_0 - K)^+]$$

The notation $(dc_t)(S_0 + (K - L_t)^+)$ (resp. $(dp_t)(S_0 - (K - L_t)^+))$ means $dc_t(Q)$ (resp. $dp_t(Q)$) where Q is replaced by $S_0 + (K - L_t)^+$ (resp. $S_0 - (K - L_t)^+$).

REMARK 3.6 Lévy's identity   The result (3.16) can be obtained directly by using Lévy's identity which states that the local time L_t and the running maximum $M_t - S_0$ of a Brownian motion (starting at S_0) are equal in distribution

$$\mathbb{E}[(L_T - K)^+] = \mathbb{E}[(M_T - S_0 - K)^+] = 2\mathbb{E}[(S_T - S_0 - K)^+]$$

The last identity comes from the reflection's principle for the Brownian motion (see Example 3.4). This result remains similar for Ocone's martingale by conditioning on the quadratic variation.   ⬜

PROOF   We introduce the process Y_t:

$$Y_t = (L_t - K)^+ + 21_{S_t > S_0} p_t(S_0 - (K - L_t)^+) + 21_{S_t \leq S_0} c_t(S_0 + (K - L_t)^+)$$

Note first that $Y_T = (L_T - K)^+$ as the Vanilla payoffs cancel out:

$$1_{S_T > S_0} \left(S_0 - (K - L_T)^+ - S_T \right)^+ = 0$$

$$1_{S_T \leq S_0} \left(S_T - S_0 - (K - L_T)^+ \right)^+ = 0$$

Then by applying Itô-Wentzell's formula, we get

$$dp_t(S_0 - (K - L_t)^+) = dp_t(S_0 - (K - L_t)^+) + \mathbb{E}_t[1_{S_0 - S_T > (K - L_t)^+}]1_{K > L_t}dL_t$$
$$dc_t(S_0 + (K - L_t)^+) = dc_t(S_0 + (K - L_t)^+) + \mathbb{E}_t[1_{S_T - S_0 > (K - L_t)^+}]1_{K > L_t}dL_t$$

$$(3.17)$$

Also,

$$d\langle p_t(S_0 - (K - L_t)^+), S\rangle_t = -\mathbb{E}_t[1_{S_0 - S_T > (K - L_t)^+}]d\langle S\rangle_t$$
$$d\langle c_t(S_0 + (K - L_t)^+), S\rangle_t = \mathbb{E}_t[1_{S_T - S_0 > (K - L_t)^+}]d\langle S\rangle_t \qquad (3.18)$$

In order to apply Itô's formula on Y_t, we introduce a smooth regularization of the Heaviside function $1^\epsilon(x)$ and define

$$Y_t^\epsilon = (L_t - K)^+ + 21^\epsilon(S_t - S_0)\left(p_t(S_0 - (K - L_t)^+) - c_t(S_0 + (K - L_t)^+)\right)$$
$$+2c_t(S_0 + (K - L_t)^+)$$

Below, we denote $\delta^\epsilon(x) \equiv \partial_x 1^\epsilon(x)$. Then, we get

$$\frac{dY_t^\epsilon}{2} = 1_{L_t > K}\frac{dL_t}{2} + d\left(c_t(S_0 + (K - L_t)^+)\right) \qquad (3.19)$$
$$+ 1^\epsilon(S_t - S_0)d\left(p_t(S_0 - (K - L_t)^+) - c_t(S_0 + (K - L_t)^+)\right)$$
$$+ \delta^\epsilon(S_t - S_0)\left(d\langle p_t(S_0 - (K - L_t)^+), S\rangle_t - d\langle c_t(S_0 + (K - L_t)^+), S\rangle_t\right)$$
$$+ \left(\delta^\epsilon(S_t - S_0)dS_t + \frac{1}{2}(\delta^\epsilon)'(S_t - S_0)d\langle S, S\rangle_t\right)$$
$$\left(p_t(S_0 - (K - L_t)^+) - c_t(S_0 + (K - L_t)^+)\right)$$

(i) By using Equations (3.17, 3.18), we get that the first three lines converge pathwise when $\epsilon \to 0$ to

$$1(S_t - S_0)(dp_t)(S_0 - (K - L_t)^+) + 1(S_0 - S_t)(dc_t)(S_0 + (K - L_t)^+)$$

where we have used that $\mathbb{E}_t[1_{S_0 - S_T > 0}] = \mathbb{E}_t[1_{S_T - S_0 > 0}] = \frac{1}{2}$ when $S_t = S_0$ by the conditional symmetry of S.

(ii) We will show that the last term in Equation (3.19) cancels out when $\epsilon \to 0$: First, from conditional symmetry property,

$$p_t(S_0 - (K - L_t)^+) - c_t(S_0 + (K - L_t)^+) = c_t(-S_0 + 2S_t + (K - L_t)^+)$$
$$-c_t(S_0 + (K - L_t)^+) \le 2|S_t - S_0|$$

Then pathwise $\delta^\epsilon(S_t - S_0)|S_t - S_0| \to 0$. For the second term, we take $\delta_\epsilon(x) \equiv \frac{1}{\sqrt{2\pi\epsilon}}e^{-\frac{x^2}{2\epsilon}}$. We obtain that the term

$$\frac{1}{2}(\delta^\epsilon)'(S_t - S_0)\left(p_t(S_0 - (K - L_t)^+) - c_t(S_0 + (K - L_t)^+)\right)d\langle S, S\rangle_t$$

is bounded by

$$\delta^\epsilon(S_t - S_0)d\langle S, S\rangle_t \frac{|S_t - S_0|^2}{\epsilon} = \int_{\mathbb{R}} dy \delta^\epsilon(y) \frac{y^2}{\epsilon} dL_t^{y+S_0}$$

where we have used the occupation formula. Finally from Exercise 1.32 (2) in [130], we have that $L_t^{y+S_0}$ is Holder continuous with $0 < \alpha < 1/2$ in y from which we conclude.

Putting all together, this gives that Y_t is a $\mathbb{P}$-local martingale for all $\mathbb{P} \in \mathcal{M}^c(0)$ and:

$$Y_T = Y_0 + 2 \int_0^T 1_{S_t > S_0} dp_t(S_0 - (K - L_t)^+) + 1_{S_t \le S_0} dc_t(S_0 + (K - L_t)^+)$$

□

Chapter 4

Continuous-time MOT and Skorokhod embedding

Abstract   We consider path-dependent payoffs which can depend on S_T, the running maximum, the running minimum, its local time and its variance $\langle \ln S \rangle_T$. We formulate MOT in continuous-time and link it to SEP. This enables to rederive known (optimal) solutions to SEP and deduce new ones, some valid in the multi-marginal case.

4.1   Continuous-time MOT and robust hedging

Our probabilistic setup is the same as the one introduced in Chapter 3. Recall that superscript σ on S^σ has been deleted. Additionally, for all $\mathbb{P} \in \mathcal{M}^c$, we denote:

$$\mathbb{H}^2_{\mathrm{loc}}(\mathbb{P}) \equiv \left\{ H \in \mathbb{H}^0(\mathbb{P}) : \int_0^T H_t^2 d\langle S \rangle_t < \infty, \ \mathbb{P} - \mathrm{a.s.} \right\},$$

Under the self-financing condition, for any delta process H, the portfolio value process

$$Y_t^H \equiv Y_0 + \int_0^t H_s dS_s, \ t \in [0, T], \tag{4.1}$$

is well-defined $\mathbb{P}-$a.s. for every $\mathbb{P} \in \mathcal{M}^c$, whenever $H \in \mathbb{H}^2_{\mathrm{loc}}$. In order to avoid doubling strategies, we introduce the set of admissible portfolios:

$$\mathcal{H} \equiv \left\{ H : \ H \in \mathbb{H}^2_{\mathrm{loc}} \text{ and } Y^H \text{ is a } \mathbb{P} - \text{supermartingale for all } \mathbb{P} \in \mathcal{M}^c \right\}.$$

4.1.1   Pathwise integration

The stochastic integral in (4.1) is typically defined $\mathbb{P}-$almost surely for some reference probability measure $\mathbb{P}$ under which the canonical process B is a semimartingale. We list some approaches to extend the notion of stochastic integration when reference probability measures are not specified. The main difficulty is that every two different probability measures $\mathbb{P}$ and $\mathbb{P}'$ in $\mathcal{M}^c$ can be singular to each other (consider for example $\mathbb{P} \equiv \mathbb{P}^\sigma$ and $\mathbb{P}' \equiv \mathbb{P}^{\sigma'}$ with $\sigma \neq \sigma'$), i.e., there are event sets A and $A' \in \mathcal{F}_T$ such that

$$\mathbb{P}[A] = 1, \ \mathbb{P}'[A] = 0, \text{ and } \mathbb{P}[A'] = 0, \ \mathbb{P}'[A'] = 1,$$

(1) The first approach is to use the so-called Föllmer–Itô calculus without probability [73]. The stochastic integral in (4.1) is defined by means of the pathwise Itô formula:

$$\int_0^t \partial_s u(r, S_r) dS_r \equiv u(t, S_t) - u(0, S_0) - \int_0^t \partial_r u(r, S_r) dr$$
$$- \frac{1}{2} \int_0^t \partial_s^2 u(s, S_s) d\langle S \rangle_s$$

This definition applies (1) only for those paths $\omega \in \Omega$ which have a finite quadratic variation and the hedge should be given by $H_t = \partial_s u(t, S_t)$ for some $C^{1,2}$-function $u : [0, T] \times \mathbb{R}_+ \longrightarrow \mathbb{R}$. This restricts the class of admissible hedging strategies.

(2) An alternative approach used by Dolinsky and Soner [63] is to define the stochastic integral by integration by parts:

$$\int_0^t H_s dS_s \equiv H_t S_t - H_0 S_0 - \int_0^t S_s dH_s,$$

This imposes H to be of finite total variation and the last integral is a well-defined Stieltjes integral.

(3) The next approach is the quasi-sure stochastic analysis developed by Denis and Martini [62]. Let $\mathcal{M}_{[\underline{\sigma}, \overline{\sigma}]}^c$ be the collection of all probability measures $\mathbb{P} \in \mathcal{M}^c$ with quadratic variation process $\langle S \rangle$ absolutely continuous with respect to the Lebesgue measure with density restricted to the interval $[\underline{\sigma}, \overline{\sigma}]$:

$$\underline{\sigma} dt \leq d\langle S \rangle_t \leq \overline{\sigma} dt, \ \mathbb{P} - \text{a.s.} \tag{4.2}$$

In financial terms, this corresponds to an uncertain volatility model where the volatility is unknown but bounded in the interval $[\underline{\sigma}, \overline{\sigma}]$. The quasi-sure stochastic integral introduced in Denis and Martini [62] is a stochastic integral $\int_0^t H_s dS_s$ defined simultaneously under all probability measures $\mathbb{P}$ in $\mathcal{M}_{[\underline{\sigma}, \overline{\sigma}]}^c$.

(4) Soner–Touzi–Zhang [136] introduces a new approach, closed in the spirit to the quasi-sure stochastic analysis, that does not suffer from the limitation of the quadratic variation process bounded by (4.2). They consider integrands H which are in the integrability set of all admissible probability measures.

(5) The various restrictions on the integrand H_t are solved by Nutz [116] who obtains a pathwise stochastic integral $\int_0^T H_s dS_s$ which coincides with the corresponding $\mathbb{P}$–Itô stochastic integral $\mathbb{P}$–a.s. for all probability measure $\mathbb{P} \in \mathcal{M}^c$ such that the quadratic variation $\langle S \rangle_t$ is well-defined.

(6) A last approach is provided in [14] using the pathwise Dambis–Dubins–Schwarz theorem introduced by Vovk [140, 141, 142].

4.1.2   Continuous-time MOT

Let ξ be a payoff defined as an $\mathcal{F}_T$−measurable random variable. In addition to the continuous-time trading, we assume that the investor can take static positions in Vanilla options with maturities $(t_i)_{i=1,\dots,n}$. The t_i-Vanilla defined by the payoff $\lambda_i(S_{t_i}) \in \mathrm{L}^1(\mathbb{P}^i)$ has an unambiguous market price given by $\mathbb{E}^{\mathbb{P}^i}[\lambda_i(S_{t_i})]$, see Section 2.1.1. The robust super-replication price is then defined by:

DEFINITION 4.1

$$\mathrm{MK}_n^c(\mathbb{P}^1,\dots,\mathbb{P}^n) \equiv \inf \left\{ Y_0 : \exists\, (\lambda_i \in \mathrm{L}^1(\mathbb{P}^i))_{i=1,\dots,n} \ and \ H \in \mathcal{H}, \right.$$
$$\left. \overline{Y}_T^{H,\lambda} \geq \xi, \mathbb{P} - a.s. \ for \ all \ \mathbb{P} \in \mathcal{M}^c \right\},$$

where $\overline{Y}^{H,\lambda}$ denotes the portfolio value of a self-financing strategy with continuous trading H in the underlying, and static trading $(\lambda_i)_{i=1,\dots,n}$ in the t_i-Vanillas:

$$\overline{Y}_T^{H,\lambda} \equiv Y_0 + \int_0^T H_s dS_s + \sum_{i=1}^n \lambda_i(S_{t_i}) - \sum_{i=1}^n \mathbb{E}^{\mathbb{P}^i}[\lambda_i(S_{t_i})] \qquad (4.3)$$

The investor has therefore the possibility of buying at time 0 any Vanilla with payoff $\lambda_i(S_{t_i})$ for the price $\mathbb{E}^{\mathbb{P}^i}[\lambda_i(S_{t_i})]$. $\mathrm{MK}_n^c(\mathbb{P}^1,\dots,\mathbb{P}^n)$ is an upper bound on the price of ξ necessary for absence of strong (model-independent) arbitrage opportunities: selling ξ at a higher price, the hedger could set up a portfolio with a negative initial cost and a non-negative payoff under any market scenario. Similarly, in the case of continuous-time static hedging in Vanillas, we define

DEFINITION 4.2

$$\mathrm{MK}_\infty^c((\mathbb{P}^t)_{t\in(0,T]}) \equiv \inf \left\{ Y_0 : \exists\, (\lambda(t,\cdot) \in \mathrm{L}^1(\mathbb{P}^t))_{t\in(0,T]} \ and \ H \in \mathcal{H}, \right.$$
$$\left. \overline{Y}_T^{H,\lambda} \geq \xi, \mathbb{P} - a.s. \ for \ all \ \mathbb{P} \in \mathcal{M}^c \right\},$$

where $t \mapsto \mathbb{E}^{\mathbb{P}^t}[\lambda(t, S_t)] \in \mathrm{L}^1([0,T])$ and

$$\overline{Y}_T^{H,\lambda} \equiv Y_0 + \int_0^T H_s dS_s + \int_0^T \lambda(t, S_t) dt - \int_0^T \mathbb{E}^{\mathbb{P}^t}[\lambda(t, S_t)] dt \qquad (4.4)$$

The next result gives a dual formulation of the robust superhedging:

PROPOSITION 4.1

Assume that $\sup_{\mathbb{P}\in\mathcal{M}^c} \mathbb{E}^{\mathbb{P}}[\xi^+] < \infty$. Then:

(i):

$$\mathrm{MK}_n^c(\mathbb{P}^1, \dots, \mathbb{P}^n) = \inf_{(\lambda_i \in \mathrm{L}^1(\mathbb{P}^i))_{i=1,\dots,n}} \sum_{i=1}^n \mathbb{E}^{\mathbb{P}^i}[\lambda_i(S_{t_i})]$$

$$+ \sup_{\mathbb{P} \in \mathcal{M}^c} \mathbb{E}^{\mathbb{P}}\Big[\xi - \sum_{i=1}^n \lambda_i(S_{t_i})\Big] \qquad (4.5)$$

(ii):

$$\mathrm{MK}_\infty^c((\mathbb{P}^t)_{t \in (0,T]}) = \inf_{\lambda(t,\cdot) \in \mathrm{L}^1(\mathbb{P}^t), \forall t \in (0,T]} \int_0^T \mathbb{E}^{\mathbb{P}^t}[\lambda(t, S_t)]dt$$

$$+ \sup_{\mathbb{P} \in \mathcal{M}^c} \mathbb{E}^{\mathbb{P}}\Big[\xi - \int_0^T \lambda(t, S_t)dt\Big] \qquad (4.6)$$

The proof is not reported (see e.g. [77, 86]) as we will prove these relations on a case-by-case basis. The weak duality result is:

$$\mathrm{MK}_n^c(\mathbb{P}^1, \dots, \mathbb{P}^n) \geq \inf_{(\lambda_i \in \mathrm{L}^1(\mathbb{P}^i))_{i=1,\dots,n}} \sum_{i=1}^n \mathbb{E}^{\mathbb{P}^i}[\lambda_i(S_{t_i})]$$

$$+ \sup_{\mathbb{P} \in \mathcal{M}^c} \mathbb{E}^{\mathbb{P}}\Big[\xi - \sum_{i=1}^n \lambda_i(S_{t_i})\Big] \qquad (4.7)$$

which could be derived as $\overline{Y}_T^{H,\lambda} \geq \xi, \mathbb{P} - $ a.s. for all $\mathbb{P} \in \mathcal{M}^c$ implies that

$$Y_0 \geq \mathbb{E}^{\mathbb{P}}\Big[\xi - \sum_{i=1}^n \lambda_i(S_{t_i})\Big] + \sum_{i=1}^n \mathbb{E}^{\mathbb{P}^i}[\lambda_i(S_{t_i})]$$

Taking the infimum over Y_0 and the supremum over $\mathbb{P} \in \mathcal{M}^c$ gives our weak duality result. The duality result means that formally

$$\inf_{H \in \mathcal{H}} Y_0 = \sup_{\mathbb{P} \in \mathcal{M}^c} \mathbb{E}^{\mathbb{P}}\Big[\xi - \sum_{i=1}^n \lambda_i(S_{t_i})\Big]$$

4.2   Matching marginals

Taking for granted that we can permute the supremum over $\mathbb{P} \in \mathcal{M}^c$ and the infimum over $(\lambda_i \in \mathrm{L}^1(\mathbb{P}^i))_{i=1,\dots,n}$, we get

$$\mathrm{MK}_n^c(\mathbb{P}^1, \dots, \mathbb{P}^n) = \sup_{\mathbb{P} \in \mathcal{M}^c} \inf_{(\lambda_i \in \mathrm{L}^1(\mathbb{P}^i))_{i=1,\dots,n}} \sum_{i=1}^n \mathbb{E}^{\mathbb{P}^i}[\lambda_i(S_{t_i})]$$

$$+ \mathbb{E}^{\mathbb{P}}\Big[\xi - \sum_{i=1}^n \lambda_i(S_{t_i})\Big]$$

Then, taking the infimum over $(\lambda_i \in L^1(\mathbb{P}^i))_{i=1,\ldots,n}$, we deduce

$$MK_n^c(\mathbb{P}^1,\ldots,\mathbb{P}^n) = \sup_{\mathbb{P}\in\mathcal{M}^c(\mathbb{P}^1,\ldots,\mathbb{P}^n)} \mathbb{E}^{\mathbb{P}}[\xi] \qquad (4.8)$$

where

$$\mathcal{M}^c(\mathbb{P}^1,\ldots,\mathbb{P}^n) \equiv \{\mathbb{P}\in\mathcal{M}^c \;:\; S_{t_i} \overset{\mathbb{P}}{\sim} \mathbb{P}^i, \quad \forall\, i=1,\ldots,n\}$$

Note that from inequalities (4.7) and (1.10), we can prove that

$$MK_n^c(\mathbb{P}^1,\ldots,\mathbb{P}^n) \geq \sup_{\mathbb{P}\in\mathcal{M}^c(\mu)} \mathbb{E}^{\mathbb{P}}[\xi] \qquad (4.9)$$

Similarly, taking for granted that we can permute the supremum over $\mathbb{P}\in\mathcal{M}^c$ and the infimum over $\lambda(t,\cdot)\in L^1(\mathbb{P}^t), \forall t\in(0,T]$, we get

$$MK_\infty^c((\mathbb{P}^t)_{t\in(0,T]}) = \sup_{\mathbb{P}\in\mathcal{M}^c((\mathbb{P}^t)_{t\in(0,T]})} \mathbb{E}^{\mathbb{P}}[\xi] \qquad (4.10)$$

where

$$\mathcal{M}^c((\mathbb{P}^t)_{t\in(0,T]}) \equiv \{\mathbb{P}\in\mathcal{M}^c \;:\; S_t \overset{\mathbb{P}}{\sim} \mathbb{P}^t, \quad \forall\, t\in(0,T]\}$$

So providing we could justify this minimax argument, our robust superhedging is connected to a MOT: we maximize the cost $\mathbb{E}^{\mathbb{P}}[\xi]$ over the space of martingale measures with marginals $(\mathbb{P}^i)_{i=1,\ldots,n}$ (or $(\mathbb{P}^t)_{t\in(0,T]}$) and $\mathbb{P}^0 = \delta_{S_0}$. If the dual is attained (not necessarily as $\mathcal{M}^c(\mathbb{P}^1,\ldots,\mathbb{P}^n)$ is not compact), $MK_n^c(\mathbb{P}^1,\ldots,\mathbb{P}^n) = \mathbb{E}^{\mathbb{P}^*}[\xi]$ should be attained by a martingale measure $\mathbb{P}^*\in\mathcal{M}^c(\mathbb{P}^1,\ldots,\mathbb{P}^n)$. Similarly, $MK_\infty^c(\mathbb{P}^t)$ should be attained by a martingale measure in $\mathcal{M}^c((\mathbb{P}^t)_{t\in(0,T]})$. The above definitions (and results) can be generalized by replacing $\mathcal{M}^c$ by $\mathcal{M}^{\text{cadlag}}$. We will also consider $\mathcal{M}^{\text{max−continuous}}$ the subset of $\mathcal{M}^{\text{cadlag}}$ such that the running maximum is continuous.

The mathematical justification of the duality (4.8) and (4.10) has been covered by various authors under various assumptions on the canonical space Ω and on ξ. See for example [14, 20, 26, 63, 64, 65, 86, 108].

Building a martingale measure in $\mathcal{M}^c((\mathbb{P}^t)_{t\in(0,T]})$ or $\mathcal{M}^c(\mathbb{P}^1,\ldots,\mathbb{P}^n)$, not necessarily optimal for a specific payoff, is already a difficult task and, although only few examples are known, this issue has generated a lot of research in probability. In mathematical finance layman's terms, this measure is connected to an arbitrage-free model calibrated exactly to the full (or discrete) implied volatility. Below, we review some classical constructions.

4.2.1 Bass's construction

We provide an example $\mathbb{P}\in\mathcal{M}^c(\mathbb{P}^1)$. We set $S_t = f(t,B_t)$ where B_t is a Brownian motion. We impose that the function f at $t=t_1$ satisfies

$$\mathbb{E}^{\mathbb{P}^1}[\mathbf{1}_{S_{t_1}<f(t_1,K)}] = N(K)$$

with N the cumulative distribution of a Gaussian with variance t_1. This implies that $f(t_1, \cdot)$ is a monotone function and $S_{t_1} \sim \mathbb{P}^1$ as for all $K \in \mathbb{R}_+$,

$$\mathbb{E}[1_{S_{t_1} < K}] = \mathbb{E}[1_{B_{t_1} < f^{-1}(t_1, K)}] = N(f^{-1}(t_1, K)) = \mathbb{E}^{\mathbb{P}^1}[1_{S_{t_1} < K}]$$

In particular,

$$f(t_1, K) \text{ s.t. } N(K) = 1 + \partial_K C(t_1, f(t_1, K))$$

with $C(t_1, K) \equiv \mathbb{E}^{\mathbb{P}^1}[(S_{t_1} - K)^+]$. Then for $t < t_1$, f satisfies the heat kernel PDE:

$$\partial_t f(t, s) + \frac{1}{2} \partial_s^2 f(t, s) = 0$$

whose solution is

$$f(t, s) = \frac{1}{\sqrt{2\pi(t_1 - t)}} \int_{\mathbb{R}} f(t_1, y) e^{-\frac{(y-s)^2}{2(t_1 - t)}} \, dy$$

This ensures that S_t is a martingale. In the financial industry, this construction is known as the "mapping".

4.2.2 Local variance Gamma model

A second example, introduced by P. Carr [39], allows to build a martingale measure matching only discrete marginals $(\mathbb{P}^i)_{i=1,\ldots,n}$ - not continuous time marginals: $\mathbb{P} \in \mathcal{M}^c(\mathbb{P}^1, \ldots, \mathbb{P}^n)$. We explain the construction for two marginals $\mathbb{P}^1$ and $\mathbb{P}^2$, the general case being obtained by iterating this construction. This example is provided by the local variance Gamma model in which the process S_t is defined as a time-homogeneous one-dimensional Itô diffusion X_t subordinated by an independent Gamma process Γ_t [39]:

$$S_t \equiv X_{\Gamma_{t-t_1}}, \quad \forall \, t > t_1$$
$$dX_t = \sigma(X_t) dB_t, \quad X_0 \sim \mathbb{P}^1$$

The distribution of the Gamma process at time t is a Gamma distribution with density:

$$\mathbb{P}\{\Gamma_t \in ds\} = \frac{s^{\frac{t}{t^*} - 1} e^{-s/t^*}}{(t^*)^{\frac{t}{t^*}} \Gamma\left(\frac{t}{t^*}\right)} ds, \quad s > 0$$

for $t^* \equiv (t_2 - t_1)$. The time-homogeneous local volatility $\sigma(\cdot)$ is defined as

$$\frac{1}{2} \sigma(K)^2 \partial_K^2 C(t_2, K) = \frac{C(t_2, K) - C(t_1, K)}{t_2 - t_1}$$

where $C(t_i, K) = \mathbb{E}^{\mathbb{P}^i}[(S_{t_i} - K)^+]$ and depends only on market values of t_1, t_2-call options. Here the volatility $\sigma(x)$ is well-defined if $\mathbb{P}^1 \leq \mathbb{P}^2$ in the convex order and if we assume the additional condition $\partial_K^2 C(t_2, K) > 0$. As a consequence of our construction, we obtain

PROPOSITION 4.2
$S_{t_i} \sim \mathbb{P}^i$, $i = 1, 2$.

PROOF
(i): The first statement $S_{t_1} = X_{t_1} \sim \mathbb{P}^1$ is trivial as $\Gamma_0 = 0$.
(ii): We have

$$p_S(t_2, x|y) = \int_0^\infty p_X(s, x|y) p(\Gamma_{t^*} = s) ds$$

with $\mathbb{P}(S_{t_2} = x | S_{t_1} = y) \equiv p_S(t_2, x|y)$ and $\mathbb{P}(X_s = x | X_0 = y) \equiv p_X(s, x|y)$. As p_X is solution of the Fokker–Planck PDE, i.e., $\mathcal{L}_x^\dagger p_X(s, x|y) = \partial_s p_X(s, x|y)$ with $\mathcal{L}_x^\dagger \cdot = \frac{1}{2} \partial_x^2 (\sigma(x)^2 \cdot)$, we deduce that

$$\mathcal{L}_x^\dagger p_S(t_2, x|y) = \int_0^\infty \mathcal{L}_x^\dagger p_X(s, x|y) p(\Gamma_{t^*} = s) ds$$

$$= \int_0^\infty \partial_s p_X(s, x|y) p(\Gamma_{t^*} = s) ds$$

$$= -\int_0^\infty p_X(s, x|y) \partial_s p(\Gamma_{t^*} = s) ds - \delta(x - y) p(\Gamma_{t^*} = 0)$$

$$= \frac{1}{t^*} (p_S(t_2, x|y) - \delta(x - y))$$

In the last line, we have used that $-\partial_s p(\Gamma_{t^*} = s) = (1/t^*) p(\Gamma_{t^*} = s)$ and $p(\Gamma_{t^*} = 0) = (1/t^*)$. Finally, as

$$\partial_x^2 C(t_2, x) = \int p_S(t_2, x|y) \mathbb{P}^1(dy)$$

we obtain

$$\frac{1}{2} \partial_x^2 \left(\sigma^2(x) \partial_x^2 C(t_2, x) \right) = \left(\partial_x^2 C(t_2, x) - \partial_x^2 C(t_1, x) \right) / t^*$$

which implies by integration our result. □

This construction can then be iterated for building a martingale measure calibrated to discrete marginals $(\mathbb{P}^i)_{i=1,\dots,n}$. As a corollary of this construction, we get

COROLLARY 4.1
The convex set $\mathcal{M}^c(\mathbb{P}^1, \dots, \mathbb{P}^n)$ is non-empty if and only if $(\mathbb{P}^i)_{i=1,\dots,n}$ are in convex order.

PROOF

$\Longrightarrow$: Jensen's inequality as for all $\mathbb{P} \in \mathcal{M}^c(\mathbb{P}^1, \ldots, \mathbb{P}^n)$,

$$\mathbb{E}^{\mathbb{P}^i}[(S_{t_i} - K)^+] = \mathbb{E}^{\mathbb{P}}[(S_{t_i} - K)^+] \geq \mathbb{E}^{\mathbb{P}}[(S_{t_{i-1}} - K)^+] = \mathbb{E}^{\mathbb{P}^{i-1}}[(S_{t_{i-1}} - K)^+]$$

$\Longleftarrow$: above construction (note that this requires the additional assumption $\partial_K^2 C(t_i, K) > 0$). $\qquad \square$

4.2.3   Local volatility model

An example of $\mathbb{P} \in \mathcal{M}^c((\mathbb{P}^t)_{t \in (0,T]})$, commonly used by practitioners in quantitative finance in part due to this property, is the local volatility model introduced by B. Dupire [68] and defined by a one-dimensional Markovian SDE:

$$dS_t = \sigma_{\mathrm{loc}}(t, S_t) dB_t, \quad \sigma_{\mathrm{loc}}(t, K)^2 \equiv 2 \frac{\partial_t C(t, K)}{\partial_K^2 C(t, K)} \qquad (4.11)$$

where $C(t, K) = \mathbb{E}^{\mathbb{P}^t}[(S_t - K)^+]$ is the (market) value of a call option with maturity t and strike K. Here the volatility $\sigma(t, s)$ is well-defined if $(\mathbb{P}^t)_{t \in (0,T]}$ are in convex order and if we assume the additional condition $\partial_K^2 C(t, K) > 0$ for all $t \in (0, T]$. By construction,

PROPOSITION 4.3
$S_t \sim \mathbb{P}^t$ *for all* $t \in (0, T]$.

PROOF   A straightforward application of Tanaka's lemma on $(S_t - K)^+$ gives

$$d(S_t - K)^+ = 1_{S_t \geq K} dS_t + \frac{1}{2} dL_t^K$$

where L_t^K is the local time at K. Formally, $dL_t^K = \delta(S_t - K)\sigma(t, K)^2 dt$. Taking the expectation on both sides, we get

$$\partial_t \mathbb{E}^{\mathbb{P}}[(S_t - K)^+] = \frac{1}{2}\sigma(t, K)^2 \partial_K^2 \mathbb{E}^{\mathbb{P}}[(S_t - K)^+]$$

By uniqueness of this PDE, we get $\mathbb{E}^{\mathbb{P}}[(S_t - K)^+] = \mathbb{E}^{\mathbb{P}^t}[(S_t - K)^+]$ for all $K \in \mathbb{R}_+$. This implies $S_t \sim \mathbb{P}^t$. $\qquad \square$

As a corollary of this explicit construction, we get a new proof of the Kellerer theorem (see [99] for technical details).

COROLLARY 4.2
The convex set $\mathcal{M}^c((\mathbb{P}^t)_{t \in (0,T]})$ *is non-empty if and only if* $(\mathbb{P}^t)_{t \in (0,T]}$ *are in convex order.*

4.2.4 Local stochastic volatility models and McKean SDEs

A natural extension of such a LV model is given by the so-called local stochastic volatility model in which S_t satisfies

$$dS_t = \sigma_t dB_t$$

where $\sigma_t = \sigma(t, S_t)a_t$ with a_t a (possibly multi-dimensional) Itô diffusion. Following closely our proof of Proposition 4.3, we can show that $S_t \sim \mathbb{P}^t$ if and only if

$$\mathbb{E}[\sigma_t^2 | S_t = S] = \sigma_{\text{loc}}(t, S)^2 \qquad (4.12)$$

where $\sigma_{\text{loc}}(t, S)$ is defined by (4.11) which depends only on the marginals $(\mathbb{P}^t)_{t \in (0,T]}$. As $\mathbb{E}[\sigma_t^2 | S_t = S] = \sigma(t, S)^2 \mathbb{E}[a_t^2 | S_t = S]$, we obtain

$$\sigma(t, S)^2 = \frac{\sigma_{\text{loc}}(t, S)^2}{\mathbb{E}[a_t^2 | S_t = S]}$$

and S_t follows a so-called non-linear McKean SDE:

$$dS_t = \frac{\sigma_{\text{loc}}(t, S_t)}{\sqrt{\mathbb{E}[a_t^2 | S_t]}} a_t dB_t \qquad (4.13)$$

The basic prototype for a McKean equation is given by the simplest McKean–Vlasov SDE, where

$$dX_t = \sigma\left(t, X_t, \mathbb{P}_t\right) dB_t \qquad (4.14)$$

$$\sigma\left(t, x, \mathbb{P}_t\right) \equiv \int \sigma(t, x, y)\mathbb{P}_t(dy) = \mathbb{E}\left[\sigma(t, x, X_t)\right]$$

Here the volatility is just the mean value $\int \sigma(t, x, y)\mathbb{P}_t(dy)$ of some function $\sigma(t, x, y)$ with respect to the distribution $\mathbb{P}_t$ of X_t.

For completeness, we cite the following theorem that gives uniqueness and existence for equation (4.14) if the volatility coefficient is Lipschitz-continuous function of x (with a linear growth condition) *and* $\mathbb{P}_t$, with respect to the so-called Wasserstein distance:

THEOREM 4.1 [115]
Let $\sigma : \mathbb{R}_+ \times \mathbb{R} \times \mathcal{P}_2(\mathbb{R}) \to \mathbb{R}$ be Lipschitz continuous function and satisfy a linear growth condition

$$|\sigma(t, X, \mathbb{P}) - \sigma(t, Y, \mathbb{Q})| \leq C\left(|X - Y| + W_2(\mathbb{P}, \mathbb{Q})\right)$$

$$|\sigma(t, X, \mathbb{P})| \leq C\left(1 + |X|\right)$$

for the sum of the canonical metric on $\mathbb{R}$ and the Monge–Kantorovich distance W_2 on the set $\mathcal{P}_2(\mathbb{R})$ of probability measures with finite second order moment:

$$W_2\left(\mu, \nu\right)^2 \equiv \inf_{\substack{\mathbb{P} \in \mathcal{P}(\mathbb{R} \times \mathbb{R}) \text{ with marginals } \mu \text{ and } \nu}} \mathbb{E}^{\mathbb{P}}[|X - Y|^2]$$

C is a positive constant. Then the nonlinear SDE

$$dX_t = \sigma(t, X_t, \mathbb{P}_t)dB_t, \qquad X_0 \in \mathbb{R}^n \qquad (4.15)$$

where $\mathbb{P}_t$ denotes the probability distribution of X_t admits a unique solution such that $\mathbb{E}(\sup_{0 \leq t \leq T} |X_t|^2) < \infty$.

Surprisingly, note that OT appears naturally - through the Wasserstein distance (i.e., MK$_2$ for the quadratic cost function on $\mathbb{R}$) - when discussing this mathematical object. Unfortunately, as the volatility $\frac{\sigma_{\mathrm{loc}}(t, S_t)}{\sqrt{\mathbb{E}[a_t^2 | S_t]}}$ does not satisfy the assumptions in the above theorem, existence and uniqueness for such a non-linear SDE are not at all obvious and still open (see e.g. Chapter 10 in [87]).

The numerical resolution of SDE (4.13) can be done with a particle's method (see e.g. Chapter 11 in [87]). The convergence of this algorithm can also be quantified with the help of OT (see Problem 15: convergence estimates in a mean-field limit in [139]). The use of (M)OT in mathematical finance seems everywhere!

4.2.5   Local Lévy's model

As an example of $\mathbb{P} \in \mathcal{M}^{\mathrm{cadlag}}((\mathbb{P}^t)_{t \in (0, T]})$, we introduce a local Lévy's model in which the process S_t is a compensated jump martingale

$$dS_t = \int_{\mathbb{R}} S_{t-} (e^x - 1) (N(dx, dt) - \nu(dx, dt))$$

where $N(dx, dt)$ is the counting measure associated with the jumps and with compensator $\nu(dx, dt) = a(t, S_t)k(x)dx dt$. This means that at each jump

$$\Delta S_t = S_{t-} e^Z$$

where Z is a random variable independent of S. with probability density $k(x)dx$. The time of default τ (after t) can be simulated as

$$\tau = \inf\{s > t : \int_t^s a(s, S_s)ds > -\ln U\}$$

where U is a uniform random variable in $[0, 1]$. In the next proposition, we prescribe a such that $\mathbb{P} \in \mathcal{M}^{\mathrm{cadlag}}((\mathbb{P}^t)_{t \in (0, T]})$:

PROPOSITION 4.4 [42]
Set $C(t, K) = \mathbb{E}^{\mathbb{P}^t}[(S_t - K)^+]$ and $a(t, s)$ the unique solution of

$$\partial_t C(t, K) = \int_0^\infty \partial_y^2 C(t, y)ya(t, y)\psi\left(\ln\left(\frac{K}{y}\right)\right) dy \qquad (4.16)$$

with $\psi(z) = \int_{\mathbb{R}} dx k(x) \left((e^x - e^z)^+ - (1 - e^z)^+ - 1_{z<0}(e^x - 1) \right)$.

Then, $S_t \overset{\mathbb{P}}{\sim} \mathbb{P}^t$ for all t.

PROOF   By applying Tanaka's lemma on $(S_t - K)^+$:

$$d(S_t - K)^+ = \int_{\mathbb{R}} \left((S_{t-}e^z - K)^+ - (S_{t-} - K)^+ \right) (N(dz, dt) - k(z)dz a_t dt)$$

$$+ \int_{\mathbb{R}} \left((S_{t-}e^z - K)^+ - (S_{t-} - K)^+ - 1_{S_{t-}>K}(e^z - 1) \right) k(z)dz a_t dt$$

with $a_t \equiv a(t, S_t)$. Taking expectation on both sides of the above equality, we get Equation (4.16). This equation admits a unique solution a as it can be written as a convolution equation:

$$\partial_t C(t, e^k) = \psi \star \left((\partial_z^2 - \partial_z)C(t, e^z)a(t, e^z) \right)(k)$$

An explicit expression for $z \mapsto (\partial_z^2 - \partial_z)C(t, e^z)a(t, e^z)$ can be obtained by taking the Fourier transform of this equation. ⧠

A general forward partial integro-differential equation for prices of call options, in a model where the dynamics of the underlying asset under the pricing measure is described by a –possibly discontinuous– semimartingale, is derived in [22] (see Theorem 2.3).

4.2.6   Martingale Fréchet–Hoeffding solution

In this section, we will assume that

Assumption 6
(1) *The marginals $\mathbb{P}^t$ are absolutely continuous with respect to the Lebesgue measure. We denote $f(t, \cdot)$ (resp. $F(t, \cdot)$) the continuous strictly positive density (resp. cumulative distribution) of S_t at time t implied from t-Vanilla option prices.*
(2) *We will assume also that there exists a unique maximizer of $\partial_t F(t, \cdot)$ that we denote $m(t)$.*

Using the market prices $C(t, K)$ of call options with strike K and maturity t, we can infer the market marginals $(\mathbb{P}^t)_{t \in (0,T]}$ by $\mathbb{P}^t(dx) = \partial_K^2 C(t, x)dx$. In practice, the Black–Scholes implied volatility $\sigma_{\mathrm{BS}}(t, K)$ of strike K and maturity $t \in (t_1, t_2)$ is not quoted and must be interpolated from the liquid maturities t_1, t_2 using for example

$$\sigma_{\mathrm{BS}}(t, K)^2 t = \left(\sigma_{\mathrm{BS}}(t_2, K)^2 t_2 - \sigma_{\mathrm{BS}}(t_1, K)^2 t_1 \right) \left(\frac{t - t_1}{t_2 - t_1} \right) + \sigma_{\mathrm{BS}}(t_1, K)^2 t_1$$

This linear interpolation guarantees that the convex order property is satisfied: $C(t_1, K) \le C(t, K) \le C(t_2, K)$ for all K.

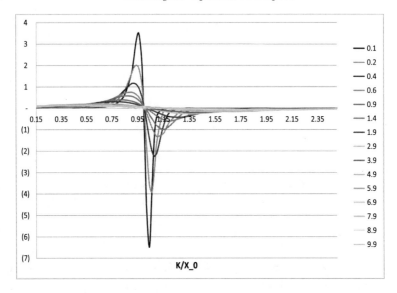

Figure 4.1: Market $\partial_t F(t, K) = \partial_t \partial_K C(t, K)$ for different (liquid) maturities t ranging from 0.1 years to 9.9 years inferred from DAX index (2-Feb-2013) as a function of K/S_0. For each t, $\partial_t F(t, \cdot)$ admits only one local maximizer. Assumption 6 is therefore satisfied on this example.

In Figure 4.1, we have plotted market $\partial_t F(t, K) = \partial_t \partial_K C(t, K)$ for different maturities t and checked that $\partial_t F$ admits only one local maximizer. Our previous assumption is therefore reasonable.

Heuristic derivation and proof

In Section 2.2.9, we have built a probability measure $\mathbb{P}^* \in \mathcal{M}(\mathbb{P}^1, \ldots, \mathbb{P}^n)$. By taking the limit $\Delta t \equiv t_i - t_{i-1} \to 0$ for all $i = 1, \ldots, n$, we will obtain a probability measure $\mathbb{P}^{\mathrm{HF}} \in \mathcal{M}^{\mathrm{cadlag}}((\mathbb{P}^t)_{t \in (0, T]})$.

$\mathbb{P}^*$ depends on the maps $(T_d^i, T_u^i)_{i=1,\ldots,n}$ solution of ODEs (2.31-2.32). In the continuous-time limit $\Delta t \to 0$, one will guess that the solution of ODEs (2.31-2.32) can be written at the first order in Δt for $s > m(t)$ as

$$T_u(t, s) = s + j_u(t, s)\Delta t$$
$$T_d(t, s) = s - j_d(t, s)$$

with $j_d(t, s) = j_u(t, s) = 0$, $\quad \forall\, s \leq m(t)$. Plugging this expression into the

ODE system, we get for all $s > m(t)$ at the first-order in Δt:

$$1 - \partial_s j_d(t, s) = -\frac{j_u(t, s)}{j_d(t, s)} \frac{f(t, s)}{\partial_t f(t, s - j_d(t, s))} \tag{4.17}$$

$$\partial_s (j_u(t, s) f(t, s)) + \partial_t f(t, s) = -\frac{j_u(t, s)}{j_d(t, s)} f(t, s) \tag{4.18}$$

Equivalently, these equations read as a Fokker–Planck-type equation satisfied by the density f:

$$\partial_t f(t, s) = -\frac{j_u(t, \underline{s})}{j_d(t, \underline{s})} \frac{f(t, \underline{s})}{1 - \partial_s j_d(t, \underline{s})} 1_{s < m(t)} - \frac{j_u(t, s)}{j_d(t, s)} f(t, s) 1_{s > m(t)}$$
$$- \partial_x (j_u(t, s) f(t, s)) 1_{s > m(t)} \tag{4.19}$$

with $\underline{s}$ the unique solution of $\underline{s} - j_d(t, \underline{s}) = s$ (here $s < m(t)$). The infinitesimal generator appearing in the above Fokker–Planck equation is obtained from a compensated jump (Markovian) martingale with compensator

$$\nu(dt, dz) = \delta(z + j_d(t, S_{t-})) \frac{j_u(t, S_{t-})}{j_d(t, S_{t-})} 1_{S_{t-} > m(t)} dz \, dt \tag{4.20}$$

where j_u, j_d are fixed by Equations (4.17, 4.18) or equivalently by Equation (4.19).

The dynamics for the stock price is given by

$$dS_t = \int_{\mathbb{R}} z \left(N(dt, dz) - \nu(dt, dz) \right) \tag{4.21}$$

where $N(dt, dz)$ is the counting measure associated with the (downwards) jumps. This means that at each jump

$$\Delta S_t = -j_d(t, S_{t-})$$

and the time of default τ (after t) can be simulated as

$$\tau = \inf\{s > t : \int_t^s \frac{j_u(s, S_{s-})}{j_d(s, S_{s-})} 1_{S_{s-} > m(s)} ds > -\ln U\}$$

where U is a uniform random variable in $[0, 1]$. This jump process corresponds to a local Lévy model as described in the previous section. We obtain

PROPOSITION 4.5
$S_t \overset{\mathbb{P}^{HF}}{\sim} \mathbb{P}^t$ for all $t \in (0, T]$.

PROOF By applying Tanaka's lemma, we get

$$d(S_t - s)^+ = 1_{S_t > s} dS_t$$
$$+ \sum_{\Delta S_{t-} \neq 0} \left((S_{t-} + \Delta S_{t-} - s)^+ - (S_{t-} - s)^+ - 1_{S_{t-} > s} \Delta S_{t-} \right)$$

By taking expectations on both sides, we obtain

$$\partial_t C(t,s) = \mathbb{E}[((S_{t-} - j_d(t, S_{t-}) - s)^+ - (S_{t-} - s)^+ + 1_{S_{t-} > s} j_d(t, S_{t-}))$$
$$\frac{j_u(t, S_{t-})}{j_d(t, S_{t-})} 1_{S_{t-} > m(t)}]$$

Finally, by differentiating twice this equation with respect to s and using that $f(t,s) = \partial_s^2 C(t,s)$, we reproduce Equation (4.19). $\qquad\square$

Optimality

By taking the limit $n \to \infty$ of Theorem 2.11, we can show that this martingale Fréchet–Hoeffding solution $\mathbb{P}^{\mathrm{HF}}$ is the optimal upper bound for a class of path-dependent payoffs characterized as

$$C(S.) \equiv \frac{1}{2} \int_0^T c_{yy}(S_t, S_t) d\langle S \rangle_t^c + \sum_{0 \le t \le T} c(S_t^-, S_t) \qquad (4.22)$$

for all $S.$ in Ω^{cadlag}. $\langle S \rangle_t^c$ means the continuous part of the quadratic variation. The function $c : \mathbb{R}_+^2 \to \mathbb{R}$ is in C^3 and satisfies

$$c(x, x) = c_y(x, x) = 0 \text{ and } c_{xyy}(x, y) > 0$$

We have

THEOREM 4.2 see [96] for details
$\mathrm{MK}_\infty^{\mathrm{cadlag}}((\mathbb{P}^t)_{t \in (0,T]})$ *for the class of payoffs defined by (4.22) is solved by* $\mathbb{P}^{\mathrm{HF}}$ *(4.21). Moreover, we have*

$$\mathrm{MK}_\infty^{\mathrm{cadlag}}((\mathbb{P}^t)_{t \in (0,T]}) = \int_0^T \int_{m(t)}^\infty \frac{j_u(t,s)}{j_d(t,s)} c(s, s - j_d(t,s)) f(t,s) \, ds \, dt.$$

As an application, let us finally consider the reward function $c(x, y) \equiv (\ln x - \ln y)^2$, corresponding to the payoff of a variance swap. More precisely, the payoff of variance swap is given by $\sum_{k=0}^{n-1} \ln^2 \frac{S_{t_{k+1}}}{S_{t_k}}$ in the discrete-time case, and by

$$\int_0^T \frac{d\langle S \rangle_t^c}{S_t^2} + \sum_{0 < t \le 1} \ln^2 \frac{S_t}{S_{t-}}$$

in the continuous-time case and belongs to the class defined above. Note that if we assume that S_t is a continuous process, we have seen that the variance swap can be replicated by holding a static position in an European log contract with payoff $-2 \ln S_T/S_0$ at the maturity $t_n \equiv T$ and a Delta hedging $H_t = 2/S_t$ (see Section 3.2). The story is different if we assume that the process S_t can jump. In this case, the replication is no more applicable and one needs to consider robust sub- and super-replication prices. Then from Theorem 4.2, the optimal upper bound of the variance swap is given by

$$\int_0^T dt \int_{m(t)}^\infty ds \; \frac{j_u(t,s)}{j_d(t,s)} \; \ln^2 \frac{s - j_d(t,s)}{s} \; f(t,s)$$

A similar construction can be obtained for the anti-monotone rearrangement martingale $\mathbb{P}^{HF}$ which corresponds to compensated upward jumps. The optimal lower bound is given by

$$\int_0^T dt \int_{\bar{m}(t)}^\infty ds \; \frac{\bar{j}_u(t,s)}{\bar{j}_d(t,s)} \; \ln^2 \frac{s - \bar{j}_d(t,s)}{s} \; \bar{f}(t,s)$$

where $\bar{m}(t)$ is the unique maximizer of $\bar{F}(t,s) \equiv 1 - F(t,-s)$. $\bar{j}_u, \bar{j}_d$ are fixed by Equations (4.17, 4.18) with F replaced $\bar{F}$.

Fake Brownian motion

These two jump processes $\mathbb{P}^{HF}$ and $\mathbb{P}^{HF}$ give rise to two new examples of (non-continuous) fake Brownian motions (if we take for f_t the density of a Brownian motion). A fake Brownian motion is a local martingale with the same marginals as a Brownian motion. We also refer to [3, 72, 90, 97, 102, 122] for other solutions and related results.

Let $\mathbb{P}^t \equiv \mathcal{N}(0,t)$. By direct computation, we have $m(t) = -\sqrt{t}$. In this case, it follows from Equations (4.17,4.18), or equivalently (2.33), that $T_d(t,s)$ is defined by the equation:

$$\int_{T_d(t,s)}^s (s - \xi)(\xi^2 - t)e^{-\xi^2/2t}d\xi = 0 \text{ for all } s \geq m(t).$$

By direct change of variables, this provides the scaled solution $T_d(t,s) \equiv t^{1/2}\widehat{T}_d(t^{-1/2}s)$, where:

$$\widehat{T}_d(s) \leq -1 \text{ is defined for all } s \geq -1 \text{ by } \int_{\widehat{T}_d(s)}^s (s - \xi)(\xi^2 - 1)e^{-\xi^2/2}d\xi = 0.$$

i.e.,

$$e^{-\widehat{T}_d(s)^2/2} \left(1 + \widehat{T}_d(s)^2 - s\widehat{T}_d(s)\right) = e^{-s^2/2}.$$

Similarly, we see that $j_u(t,s) \equiv t^{-1/2}\widehat{j}_u(t^{-1/2}s)$, where for all $s \geq -1$,

$$\widehat{j}_u(s) \equiv \frac{1}{2}[s - \widehat{T}_d(s)e^{-(\widehat{T}_d(s)^2-s^2)/2}] = \frac{1}{2}\left[s - \frac{\widehat{T}_d(s)}{1 + \widehat{T}_d(s)^2 - s\widehat{T}_d(s)}\right]$$

We have plotted the maps $\widehat{T}_d(s)$ and $\widehat{T}_u(s) \equiv s + \widehat{j}_u(s)$ in Figure 4.2. A similar construction can be achieved for self-similar marginals (see [96]). From Theorem 4.2, our fake Brownian maximizes the expectation of the quadratic variation within the class $\mathcal{M}^{\text{cadlag}}((\mathbb{P}^t)_{t\in(0,T]})$.

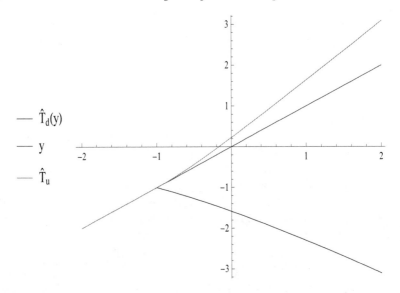

Figure 4.2: Fake Brownian motion: Maps $\hat{T}_d$ and $\hat{T}_u$ (top curve). Note the similarity with the discrete martingale Fréchet–Hoeffding solution - see Figure 2.4.

4.3   Digression: Matching path-dependent options

In the previous section, we have listed examples of some arbitrage-free models calibrated on Vanillas. Although Vanillas are the most liquid instruments, they are however not always suitable hedging instruments for exotic options depending on forward implied volatility dynamics. In this section, we consider continuous diffusions, specified by $dS_t = \sigma_t dB_t$, where the stochastic volatility σ_t is chosen such that we match market prices of path-dependent options. As an example, we consider barrier option $C(t, K, B) \equiv \mathbb{E}[(S_t - K)^+ 1_{M_t > B}]$ depending on the running maximum $M_t = \max_{s \leq t} S_s$. The next proposition generalises the condition (4.12) which implies that our diffusive model is calibrated to Vanillas.

PROPOSITION 4.6
(i) *Barrier option prices $C(t, K, B) \equiv \mathbb{E}[(S_t - K)^+ 1_{M_t > B}]$ for all (B, K, t) are matched if and only if*

$$\mathbb{E}[\sigma_t^2 | S_t = K, M_t = B] = 2 \frac{\partial_{Bt}\left(C(t, K, B) + \{(B - K)^+ \partial_K C(t, K, B)|_{K=0}\}\right)}{\partial_{KKB} C(t, K, B)}$$

A similar formula appears in [89].

(ii) *Options on variance* $C(t, K) \equiv \mathbb{E}[(\langle \ln S \rangle_t - K)^+]$ *for all* (K, t) *are matched if and only if*

$$\mathbb{E}[\frac{\sigma_t^2}{S_t^2} | \langle \ln S \rangle_t = K] = -\frac{\partial_{tK} C(t, K)}{\partial_{K^2} C(t, K)}$$

(iii) *Options on Leverage ETF,* $C(t, K) \equiv \mathbb{E}[(l_t - K)^+]$ *for all* (K, t) *with* $l_t \equiv l_0 \left(\frac{S_t}{S_0}\right)^\beta e^{-\frac{1}{2}(\beta^2 - \beta)\langle \ln S \rangle_t}$, *are matched if and only if*

$$\mathbb{E}[\sigma_t^2 | L_t = K] = \frac{2\partial_t C(t, K)}{\beta^2 \partial_{K^2} C(t, K)}$$

(iv) *Asian options,* $C(t, K, B) \equiv \mathbb{E}[(S_t - K)^+ 1_{A_t > B}]$ *for all* (B, K, t) *with* $A_t = \int_0^t S_s ds$, *are matched if and only if*

$$\mathbb{E}[\sigma_t^2 | S_t = K, A_t = B] = 2 \frac{\partial_{Bt} C(t, K, B) + \partial_B^2 \mathbb{E}^{\mathbb{P}^{mkt}} [1_{A_t > B} (S_t - K)^+ S_t]}{\partial_{KKB} C(t, K, B)}$$

where

$$\mathbb{E}^{\mathbb{P}^{mkt}} [1_{A_t > B} (S_t - K)^+ S_t] = 2 \int_K^\infty C(t, x, B) dx + KC(t, K, B)$$

PROOF   We prove only (i) as the other results use similar arguments. By applying Tanaka's lemma on $(S_t - K)^+ 1_{M_t > B}$, we get

$$d\left((S_t - K)^+ 1_{M_t > B}\right) = \left(1_{S_t > K} dS_t + \frac{1}{2}\sigma_t^2 dt\delta(S_t - K)\right) 1_{M_t > B} dt$$
$$+ (S_t - K)^+ \delta(M_t - B) dM_t$$

Then, we use that $(S_t - K)^+ \delta(M_t - B) dM_t = (M_t - K)^+ \delta(M_t - B) dM_t = (B - K)^+ \delta(M_t - B) dM_t$. Taking the expectation on both sides and differentiating with respect to B, we obtain

$$\partial_{Bt} C(t, K, B) = \frac{1}{2}\partial_{KKB} C(t, K, B) \mathbb{E}[\sigma_t^2 | S_t = K, M_t = B]$$
$$+ \partial_B \{(B - K)^+ \partial_t \mathbb{E}[1_{M_t > B}]\}$$

As $\partial_K C(t, K, B)|_{K=0} = -\mathbb{E}[1_{M_t > B}]$, we get our result.   □

4.4   Link with Skorokhod embedding problem

If we assume that the payoff $\xi(S_T, M_T, m_T, \langle S \rangle_T)$ depends on the spot S_T, the running maximum M_T, minimum m_T or the quadratic variation $\langle S \rangle_T$ at

T , by doing a stochastic time change (see DDS Theorem 3.3), $\mathrm{MK}_1^c(\mu)$, as given by formula (4.8), can be framed as a constrained perpetual American options:

$$\mathrm{MK}_1^c(\mu) = \sup_{\tau \in \mathcal{T} \,:\, B_\tau \sim \mu} \mathbb{E}^{\mathbb{P}}\left[\xi\left(B_\tau, \max_{0 \le s \le \tau} B_s, \min_{0 \le s \le \tau} B_s, \tau\right)\right]$$

This problem corresponds then to the determination of an (optimal) stopping time τ^* such that $B_{\tau^*} \sim \mu$. This is a Skorokhod embedding problem (in short SEP). More precisely,

DEFINITION 4.3 Skorokhod embedding problem
Find a stopping time τ such that $B_\tau \sim \mu$ and $B^\tau \equiv (B_{t \wedge \tau})_{t \ge 0}$ is uniformly integrable.

Additionally, we impose that $B_{t \wedge \tau}$ is an uniformly integrable martingale. We will disregard this condition in the following as we have decided not to consider the distinction between local and true martingales. The uniformly integrable condition is indeed needed to prove that a delta hedging strategy $\int_0^t H_t dS_t$ is a true martingale, and not only a local martingale.

The use of SEP for deriving model-independent bounds for exotic options consistent with T-Vanillas was pioneered by D. Hobson in the case of lookback options [100] and then by B. Dupire in the case of options on variance [69]. Note that this approach should give a remarkable solution of SEP as the associated martingale S_t has the property to maximize the expectation of the payoff ξ over the class of arbitrage-free models calibrated to T-Vanillas. This link between MOT and SEP has been recently explored in [15].

Although most known solutions of SEP share this optimality property, this property was checked in a second step in the original papers on SEP. Our approach consisting in solving this problem of perpetual American options implies at the beginning that our solution to SEP satisfies an optimality property. In the next section, we show on the simple example of an increasing function $\xi = g(M_T)$ depending on the running maximum M_T how to reproduce Azéma–Yor solution [8, 9]. Although the derivation of this solution is simple using martingale techniques (mainly a stopped Azéma–Yor martingale, see definition 3.2), we highlight that our method is straightforward and can easily handle more complicated examples. The determination of a SEP boils down in the analytical resolution of a one (eventually two)-dimensional time-homogeneous obstacle problem. In particular, following the main steps sketched below (see also Section 4.6.1.2), one can show that the optimality of Azéma–Yor solution is still valid with a more general payoff $\xi = g(S_T, M_T)$ where $s \mapsto \frac{\partial_m g(s,m)}{m-s}$ is a nondecreasing function for all $s < m$ (see [94]). Our approach can also tackle the multi-marginals SEP (see [94] and Section 4.6.1.4).

4.5 A (singular) stochastic control approach

We sketch our general strategy to solve $\mathrm{MK}_1^c(\mu)$: (i) on the primal side, we compute the optimal bounds and deduce the pathwise super(sub)-replication strategy. (ii) On the dual side, we obtain the optimal martingale measure characterized as a solution to (SEP).

Our approach involves 3 steps, that we sketch in the case of a lookback payoff $g(M_T)$. Additional details will be provided in the proof of Theorem 4.4.

Step 1: Set

$$u^\lambda(s, m) \equiv \sup_{\tau \in [0,\infty]} \mathbb{E}[g(M_\tau) - \lambda(B_\tau)|S_0 = s, M_0 = m]$$

where $M_s = \max_{0 \le r \le s} B_s$ here.

For the ease of notations, we have decided to use the same notation M_t for denoting the maximum on the underlying S_t and the maximum on the Brownian motion B_t. Proper definitions can be found by looking if M_t appears with S_t or B_t.

Then, from Proposition 4.1, $\mathrm{MK}_1^c(\mu)$ can be written as

$$\mathrm{MK}_1^c(\mu) = \inf_{\lambda \in L^1(\mu)} \left\{ \mathbb{E}^\mu[\lambda(S_T)] + u^\lambda(S_0, S_0) \right\}. \tag{4.23}$$

u^λ is a viscosity solution of the variational HJB equation:

$$\max\left(g(m) - \lambda(s) - u^\lambda(s, m), \frac{1}{2}\partial_s^2 u^\lambda(s, m) \right) = 0, \quad s < m$$

$$\partial_m u^\lambda(m, m) = 0 \tag{4.24}$$

Note that there is non uniqueness result (i.e., no comparison result) for such a PDE, in particular we will derive a family of solutions.

REMARK 4.1 Formal HJB The above variational HJB could appear strange to the reader that must be instead aware of the classical HJB equation, that writes formally

$$\partial_t u(t, s, m) + \sup_{\sigma \in [0,\infty]} \frac{\sigma^2}{2}\partial_s^2 u(t, s, m) = 0, \quad u(T, s, m) = g(m) - \lambda(s)$$

$$\partial_m u(t, m, m) = 0$$

Below, we will comment that these two PDEs are formally equivalent:
(i) If we take the supremum over σ, we will get $+\infty$ if $\partial_s^2 u > 0$, for which no solution exists. We should have therefore $\partial_s^2 u \le 0$ and the HJB equation reads

$$\partial_t u(t, s, m) = 0 \tag{4.25}$$

(ii) Similarly, the terminal condition $u(T, s, m) = g(m) - \lambda(s)$ could not hold as $\partial_s^2 u(T, s, m)$ is not required to be negative except in the case where λ is convex. Due to the unbounded control, the terminal condition is *face-lifted* and $u(T, s, m)$ should be the smallest function that satisfied

$$\partial_s^2 u(T, s, m) \leq 0, \quad u(T, s, m) \geq g(m) - \lambda(s)$$

The second inequality is obvious as

$$u(t, s, m) \equiv \sup_{\sigma \in [0, \infty]} \mathbb{E}[g(M_T) - \lambda(S_T) | S_t = s, M_t = m] \geq g(M_t) - \lambda(S_t)$$

This is equivalent to the variational equation

$$\max \left(\partial_s^2 u(T, s, m), g(m) - \lambda(s) - u(T, s, m) \right) = 0$$

(iii) Note that the solution of PDE (4.25) is $u(t, s, m) = u(T, s, m)$ and we reproduce (formally) the variational HJB equation (4.24).

$\Box$

Step 2 (primal side): Find a solution $U(s, m)$. Here we have skipped the subscript λ on U. As PDE (4.24) can be interpreted as the pricing PDE for an infinite horizon American option (i.e., obstacle problem), we look for a solution of the form:

$$\frac{1}{2} \partial_s^2 U = 0, \quad \forall (s, m) \in \mathcal{D} \cap \{s \leq m\}$$
$$U(s, m) = g(m) - \lambda(s), \quad \forall (s, m) \in \mathcal{D}^c \cap \{s \leq m\}$$

with $\mathcal{D}$, a domain in the space (s, m), interpreted as a continuation region. Outside $\mathcal{D}$, we exercise the payoff and $U(s, m) = g(m) - \lambda(s)$. By using the Neumann condition $\partial_m U(m, m) = 0$, this specifies $\lambda^*(\cdot)$, once the domain $\mathcal{D}$ is given.

REMARK 4.2 Note that points (m, m) should be in $\mathcal{D}$. Indeed, if we assume that $(m, m) \notin \mathcal{D}$, the Neumann condition can not be fulfilled as $U(s, m) = g(m) - \lambda^*(s)$ (except in the trivial case where g is a constant).

$\Box$

Additionally, we impose a smooth-fit condition

$$\partial_s U(s, m) = -\lambda^*(s), \quad \forall (s, m) \in \partial \mathcal{D} \cap \{s \leq m\}$$

This corresponds to impose that the delta is continuous at the boundary. This last equation will fix the boundary $\partial \mathcal{D}$. Once such a solution is found (i.e., satisfied the variational PDE (4.24)), one deduces that

$$\mathrm{MK}_1^c(\mu) \leq U(S_0, S_0) + \mathbb{E}^\mu[\lambda^*(S_T)]$$

Indeed, by applying Itô's lemma on U, we get for all stopping time τ

$$U(B_\tau, M_\tau) = U(S_0, S_0) + \int_0^\tau \partial_s U(B_t, M_t) dB_t + \int_0^\tau \frac{1}{2} \partial_s^2 U(B_t, M_t) dt$$

$$+ \int_0^\tau \partial_m U(M_t, M_t) dM_t$$

$$\leq U(S_0, S_0) + \int_0^\tau \partial_s U(B_t, M_t) dB_t$$

as $\partial_s^2 U(B_t, M_t) \leq 0$ and $\partial_m U(M_t, M_t) = 0$ from (4.24). We have used $B_0 = M_0 = S_0$ and $M_t = \max_{0 \leq r \leq t} B_r$. Moreover as $U(B_\tau, M_\tau) \geq g(M_\tau) - \lambda^*(B_\tau)$ from PDE (4.24), this implies that for all τ

$$U(S_0, S_0) + \int_0^\tau \partial_s U(B_t, M_t) dB_t \geq g(M_\tau) - \lambda^*(B_\tau)$$

In particular, by taking the expectation on both sides, we get

$$\sup_{\tau \in [0,\infty]} \mathbb{E}[g(M_\tau) - \lambda^*(B_\tau)] \leq U(S_0, S_0) \tag{4.26}$$

By DDS theorem (or by applying directly Itô's lemma on $U(S_t, M_t)$ with $M_t = \max_{s \geq 0} S_s$ here), this is equivalent to: For all $\mathbb{P} \in \mathcal{M}^c$,

$$U(S_0, S_0) + \int_0^T \partial_s U(S_t, M_t) dS_t + \lambda^*(S_T) \geq g(M_T)$$

This is precisely our pathwise super-replication consisting of a delta-hedging with $H_t = \partial_s U(S_t, M_t)$ and a static position in a T-Vanilla with payoff $\lambda^*(S_T)$. From the Definition 4.1, this implies that

$$\mathrm{MK}_1^c(\mu) \leq U(S_0, S_0) + \mathbb{E}^\mu[\lambda^*(S_T)]$$

In practise, finding a solution requires some conditions on g (see for example the conditions in Table 4.1).

Step 3 (dual side): Find a solution of SEP of the form

$$\tau^* \equiv \inf\{t > 0 : (B_t, M_t) \notin \mathcal{D}\}$$

from which we obtain the inequality using (4.9):

$$\sup_{\mathbb{P} \in \mathcal{M}^c(\mu)} \mathbb{E}[g(M_T)] \overset{\mathrm{DDS}}{=} \sup_{\tau : B_\tau \sim \mu} \mathbb{E}[g(M_\tau)] \leq \mathrm{MK}_1^c(\mu)$$

$$\leq U(S_0, S_0) + \mathbb{E}^\mu[\lambda^*(S_T)]$$

As $B_{\tau^*} \sim \mu$ by construction, this implies that

$$\mathbb{E}[g(M_{\tau^*})] \leq \mathrm{MK}_1^c(\mu) \leq U(S_0, S_0) + \mathbb{E}^\mu[\lambda^*(S_T)]$$

We conclude the optimality of our bound by checking that

$$U(S_0, S_0) + \mathbb{E}^\mu[\lambda^*(S_T)] = \mathbb{E}[g(M_{\tau^*})]$$

Note that the use of Proposition 4.1 is not needed as one needs only to check that our pathwise super-replication strategy is valid and then use the weak duality result (4.9). The result of Proposition 4.1 is in fact implied from the inequality:

$$\mathrm{MK}_1^c(\mu) \overset{(4.7)}{\geq} \inf_{\lambda \in \mathrm{L}^1(\mu)} \left\{ \mathbb{E}^\mu[\lambda(S_T)] + u^\lambda(S_0, S_0) \right\} \overset{(4.9)}{\geq} \sup_{\mathbb{P} \in \mathcal{M}^c(\mu)} \mathbb{E}^\mathbb{P}[g(M_T)]$$

$$\geq \mathbb{E}[g(M_{\tau^*})]$$

as $\mathrm{MK}_1^c(\mu) = \mathbb{E}[g(M_{\tau^*})]$.

REMARK 4.3   (Link with timer-like options) Once the optimal SEP τ^* is found, the upper bound reads $\mathbb{E}[g(M_{\tau^*})]$. This is nothing else than a timer-like option, considered in Chapter 3, for which we have already shown that it is model-independent.   □

4.6   Review of solutions to SEP and its interpretation in mathematical finance

We review solutions of SEP, mainly Azéma–Yor [8], Perkins [127], Vallois [137], Root [131], Rost [132] and highlight their interpretation in mathematical finance (see [101, 118] for an exhaustive list of known solutions). Following the lines outlined in the previous section, we rederive these (optimal) solutions and explain how they can be extended to the multi-marginal setup. In Table 4.1, we have listed some SEP solutions. For each SEP, we give the domain, the condition satisfied by the payoff for which the SEP solution is optimal for $\mathrm{MK}_1^c(\mu)$ and the corresponding variational PDE associated to the singular stochastic control problem.

4.6.1   Azéma–Yor solution

Before using our stochastic control approach, we recall the original derivation of the Azéma–Yor solution using martingale techniques (see also [47]).

4.6.1.1   Martingale approach

PROPOSITION 4.7 Embedding [8, 9]
Define the stopping time $\tau_{\mathrm{AY}} = \inf\{t \geq 0 \ : \ B_t \leq \psi_\mu(M_t)\}$ with $\psi_\mu^{-1}(x) =$

Table 4.1: Review of solutions to SEP. $\mathcal{L}^0 \equiv \frac{1}{2}\partial_s^2 + \partial_t$ and $\mathcal{L}^1 \equiv \frac{1}{2}\partial_s^2 + \delta(s - S_0)\partial_t$.

τ_{SEP}	Payoff	PDE
Azéma–Yor	$g(m)$ incr.	$\max\left(g - \lambda - u^\lambda, \partial_s^2 u^\lambda\right) = 0, \; \partial_m u\big|_{m,m} = 0$
Root	$g(t)$ conc.	$\max\left(g - \lambda - u^\lambda, \mathcal{L}^0 u^\lambda\right) = 0$
Vallois	$g(l)$ conv.	$\max\left(g - \lambda - u^\lambda, \mathcal{L}^1 u^\lambda\right) = 0$

$\frac{\mathbb{E}^\mu[S_T 1_{S_T \geq x}]}{\mathbb{E}^\mu[1_{S_T \geq x}]}$. Then, $B_{\tau_{AY}} \sim \mu$, *i.e., solution to SEP.*

ψ_μ is an increasing function and $\psi_\mu(x) < x$. We shall prove this result under the assumptions that μ is absolutely continuous with respect to the Lebesgue measure, with connected support and positive density.

PROOF using optional sampling theorem   By applying the optional sampling theorem to the Azéma–Yor martingale $F(M_t) - (M_t - B_t)f(M_t)$ with $a^* = S_0$ and $F(x) = \int_{S_0}^x f(u)dy$ (see Proposition 3.2), we have

$$\mathbb{E}[F(M_{\tau_{AY}})] = \mathbb{E}[(M_{\tau_{AY}} - B_{\tau_{AY}})f(M_{\tau_{AY}})]$$

which is equivalent to

$$\int_{S_0}^\infty F(x)d\mathbb{P}(M_{\tau_{AY}} \geq x) = \int_{S_0}^\infty (x - \psi_\mu(x))f(x)d\mathbb{P}(M_{\tau_{AY}} \geq x)$$

By doing an integration by part, we get for all f

$$\int_{S_0}^\infty f(x)\left(\mathbb{P}(M_{\tau_{AY}} \geq x) + (x - \psi_\mu(x))\frac{d\mathbb{P}(M_{\tau_{AY}} \geq x)}{dx}\right)dx = 0$$

We deduce that

$$\frac{d\mathbb{P}(M_{\tau_{AY}} \geq x)}{dx} = \frac{\mathbb{P}(M_{\tau_{AY}} \geq x)}{\psi_\mu(x) - x}$$

As $\mathbb{P}(B_{\tau_{AY}} \geq x) = \mathbb{P}(M_{\tau_{AY}} \geq \psi_\mu^{-1}(x))$, the above equation gives

$$\frac{d\mathbb{P}(B_{\tau_{AY}} \geq x)}{\mathbb{P}(B_{\tau_{AY}} \geq x)} = \frac{d\psi_\mu^{-1}(x)}{x - \psi_\mu^{-1}(x)}$$

This can be integrated out and we get

$$\psi_\mu^{-1}(x) = \frac{\mathbb{E}[B_{\tau_{AY}} 1_{B_{\tau_{AY}} \geq x}]}{\mathbb{E}[1_{B_{\tau_{AY}} \geq x}]}$$

If we impose that $B_{\tau_{AY}} \sim \mu$, we conclude.   ⬜

PROOF using PDE   By using Lemma 3.1 with $\lambda(x) = 1_{x<K}$, we get

$$\mathbb{E}[1_{B_{\tau_{AY}}<K}] - 1 = \frac{S_0}{\psi_\mu^{-1}(K) - K} e^{-\int_{\psi_\mu(S_0)}^{K} \frac{dy}{\psi_\mu^{-1}(y)-y}}$$

where we have used that $\psi_\mu(S_0) = 0$. If we impose that $\mathbb{E}[1_{B_{\tau_{AY}}<K}] = F_\mu(K)$
(i.e., $B_{\tau_{AY}} \sim \mu$), then after a straightforward computation, this implies that
$\psi_\mu(K)$ should be equal to

$$\psi_\mu^{-1}(K) = \frac{\mathbb{E}[S_T 1_{S_T \geq K}]}{\mathbb{E}^\mu[1_{S_T \geq K}]}$$

We reproduce Azéma–Yor solution.   ⬜

THEOREM 4.3 Optimality of the Azéma–Yor solution
Let τ be a solution to SEP. Then for all increasing function g,

$$\mathbb{E}[g(M_\tau)] \leq \mathbb{E}[g(M_{\tau_{AY}})]$$

PROOF   Without loss of generality, we can assume that $B_0 = 0$ by shifting
the Brownian motion by $-B_0$ (similarly for the maximum). Take $f(x) = 1_{x \geq \alpha}$.
Azéma–Yor martingale (with $a^* = \alpha$) is

$$N_t = (B_t - \alpha)1_{M_t \geq \alpha}$$

By applying the optional sampling theorem to the martingale N_t for all τ a
solution to SEP, we have the identity

$$\alpha\mathbb{E}[1_{M_\tau \geq \alpha}] = \mathbb{E}[B_\tau 1_{M_\tau \geq \alpha}] \tag{4.27}$$

Besides we have

$$\mathbb{E}[1_{M_\tau \geq \alpha}] \geq \mathbb{E}[1_{B_\tau \geq \alpha}] \tag{4.28}$$

Set $p = \mathbb{E}[1_{M_\tau \geq \alpha}]$ and $\lambda(x) = \mathbb{E}[1_{B_\tau \geq x}]$. Note that $\lambda(\cdot)$ is a nonincreasing
continuous function. The inequality (4.28) reads

$$p \geq \lambda(\alpha)$$

Therefore $\alpha \geq \lambda^{-1}(p)$. Starting from the equality (4.27), we have the following
chain of inequalities:

$$\alpha p \overset{(4.27)}{=} \mathbb{E}[B_\tau 1_{M_\tau \geq \alpha}] \leq \mathbb{E}[B_\tau 1_{B_\tau \geq \lambda^{-1}(p)}]$$
$$= \mathbb{E}[1_{B_\tau \geq \lambda^{-1}(p)}]\psi_\mu^{-1}(\lambda^{-1}(p)) \text{ as } B_\tau \sim \mu$$
$$= p\psi_\mu^{-1}(\lambda^{-1}(p))$$

which leads to $\alpha \leq \psi_\mu^{-1}(\lambda^{-1}(p))$ (i.e., $\lambda(\psi_\mu(\alpha)) \geq p$). Finally,

$$\mathbb{E}[1_{M_\tau \geq \alpha}] = p \leq \lambda(\psi_\mu(\alpha)) = \mathbb{E}[1_{B_{\tau_{AY}} \geq \psi_\mu(\alpha)}] = \mathbb{E}[1_{M_{\tau_{AY}} \geq \alpha}]$$

This implies our result for a general increasing function g. $\qquad\square$

4.6.1.2   Stochastic control approach

The aim of this section is to reproduce with our stochastic control approach Azéma–Yor solution to the Skorokhod embedding problem which gives the optimal bound for options written on the maximum of an underlying: $\xi = g(M_T)$ with $M_T = \max_{s \leq T} S_s$. g is an increasing function. Here we do not reproduce our proof in [77] but instead sketch the derivation highlighting the main ideas, following closely Section 4.5, and the novelty of our approach. In particular, we will derive the pathwise superhedging strategy associated to the optimal bound.

THEOREM 4.4
Let $g(M)$ be some increasing C^1 function and ψ be an increasing C^1 function such that $\psi(x) < x$. The following inequality holds:
(i): For all $\mathbb{P} \in \mathcal{M}^c$,

$$-\int_0^T \lambda'(\min(\psi(M_t), S_t))dS_t + \lambda(S_t) \geq g(M_T) \tag{4.29}$$

where λ is solution of the ODE:

$$\psi'(m)\lambda''(\psi(m))(m - \psi(m)) = g'(m) \tag{4.30}$$

(ii): This inequality is attained for $\mathbb{P}^*$ defined as the distribution of $S_t = B_{\tau_{AY} \wedge \frac{t}{T-t}}$:

$$-\int_0^T \lambda_\mu'(\min(\psi_\mu(M_t), S_t))dS_t + \lambda_\mu(S_t) = g(M_T), \quad \mathbb{P}^* - a.s.$$

where $(\lambda_\mu'(\psi_\mu(m)))' = g'(m)/(m - \psi_\mu(m))$.
(iii):

$$\mathrm{MK}_1^c(\mu) = \mathbb{E}^\mu[\lambda_\mu(S_T)]$$

(iv):

$$\mathrm{MK}_1^{\text{max continuous}}(\mu) = \mathrm{MK}_1^c(\mu)$$

In $S_t = B_{\tau_{AY} \wedge \frac{t}{T-t}}$, the function $t/(T-t)$ is arbitrary and can be replaced by any increasing function $\alpha(t)$ such that $\alpha(T)$ diverges.

REMARK 4.4 The following inequality holds also: For all $\mathbb{P} \in \mathcal{M}^c$,

$$- \int_0^T \lambda'(\psi(M_t))dS_t + \lambda(S_t) \geq g(M_T) \qquad (4.31)$$

▯

PROOF
(i): Following the method outlined in Section 4.5, we look at a solution U of the variational PDE (4.24) of the form:

$$\partial_s^2 U(s,m) = 0, \quad \psi(m) < s < m$$
$$\partial_m U(m,m) = 0 : \text{ (normal reflection)}$$
$$U(\psi(m),m) = g(m) - \lambda(\psi(m)) : \text{ (instantaneous stopping)}$$

that we complete by the additional requirement (the delta is smooth at the boundary)

$$\partial_s U(\psi(m),m) = -\lambda'(\psi(m)) : \text{ (smooth fit)}$$

This can be solved explicitly:

$$U(s,m) = g(m) - \lambda(s) + \int_{\psi(m)}^{\max(s,\psi(m))} (s-x)\lambda''(x)dx \qquad (4.32)$$

In order to satisfy the obstacle problem, we should impose that λ is a convex function, λ'' is then the second derivative measure of the convex function. Due to the normal reflection condition, λ should satisfy ODE (4.30). Note that as no initial condition is specified, the solution of this ODE (and the solution of the variational equation (4.24)) is not unique. The condition λ convex implies that g should be increasing. This is the origin of our assumption on g.

From the explicit solution U, one can justify from a classical verification argument using Itô's formula that U is a super-solution of PDE (4.24). Furthermore, by applying Itô's lemma on U, we get our pathwise super-replication. Note that we have an other supersolution

$$U(s,m) = g(m) - \lambda(s) + \int_{\psi(m)}^{s} (s-x)\lambda''(x)dx \qquad (4.33)$$

This gives the pathwise inequality (4.31).
(ii), **(iii)**: The optimality is proved from the weak duality and by taking $\psi = \psi_\mu$. Note that by construction $S_T = B_{\tau_{AY}} \sim \mu$.
(iv): For all $\mathbb{P} \in \mathcal{M}^{\max-\text{continuous}}$,

$$(M_t - S_t)dM_t = 0$$

and

$$U(S_T, M_T) = U(S_0, S_0) + \int_0^T \partial_s U(S_t, M_t)dS_t + \frac{1}{2}\int_0^T \partial_s^2 U(S_s, M_s)d\langle S\rangle_t^c$$

$$+ \int_0^T \partial_m U(S_t, M_t)dM_t$$

$$+ \sum_{\Delta S_t \neq 0}(U(S_t + \Delta S_t, M_t) - U(S_t, M_t) - \partial_s U(S_t, M_t)\Delta S_t)$$

As U is concave by construction and $\partial_m U(S_t, M_t)dM_t = g'(M_t)\frac{(M_t - S_t)}{M_t - \psi(M_t)}dM_t = 0$, we get the pathwise inequality (4.29). This implies that

$$\mathrm{MK}_1^{\max-\text{continuous}}(\mu) \leq \mathrm{MK}_1^c(\mu)$$

The reverse inequality is obtained by noting that

$$\mathrm{MK}_1^{\max-\text{continuous}}(\mu) \geq \sup_{\mathbb{P}\in\mathcal{M}^{\max-\text{continuous}}(\mu)} \mathbb{E}^{\mathbb{P}}[g(M_T)] \geq \mathrm{MK}_1^c(\mu)$$

$\square$

REMARK 4.5 Azéma–Yor, again   The solution U depends on ψ, restricted to be an increasing function with $\psi(x) < x$. As being a supersolution, this implies that

$$\mathrm{MK}_1^c(\mu) \leq \inf_{\psi \text{ incr.}} U(S_0, S_0) + \mathbb{E}^{\mu}[\lambda(S_T)]$$

where λ is given by (4.30). We have

$$\mathbb{E}^{\mu}[\lambda(S_T)] = \int \lambda''(dx)C^{\mu}(x) = \int C^{\mu}(\psi(y))\lambda''(\psi(y))\psi'(y)dy$$

$$\overset{(4.30)}{=} \int C^{\mu}(\psi(y))\frac{g'(y)}{y - \psi(y)}dy$$

with $C^{\mu}(x) \equiv \mathbb{E}^{\mu}[(S_T - x)^+]$. Pointwise (unconstrained) minimisation of $y \mapsto \frac{C^{\mu}(\psi(y))}{y - \psi(y)}$ over ψ gives

$$(y - \psi(y))\partial_y C^{\mu}(\psi(y)) + C^{\mu}(\psi(y)) = 0$$

By setting $y = \psi^{-1}(x)$, we get

$$(\psi^{-1}(x) - x)\partial_x C^{\mu}(x) + C^{\mu}(x) = 0$$

for which the solution is

$$\psi^{-1}(x) = \frac{x\partial_x C^{\mu}(x) - C^{\mu}(x)}{\partial_x C^{\mu}(x)} = \frac{\mathbb{E}^{\mu}[S_T 1_{S_T \geq x}]}{\mathbb{E}^{\mu}[1_{S_T \geq x}]} \qquad (4.34)$$

This reproduces again Azéma–Yor solution. ⬛

Example 4.1 One-touch digital, $g(m) = 1_{m>B}$
By applying our Theorem 4.4, we get the optimal pathwise super-replication
(see inequality (4.31))

$$1_{M_T>B} \le \frac{(S_T - \psi(B))^+}{B - \psi(B)} - \int_0^T \frac{1_{M_t \ge B}}{B - \psi(B)} dS_t$$

where we have used that $\lambda(x) = \frac{(x - \psi(B))^+}{B - \psi(B)}$ (see Equation (4.30)). This is
equivalent to

$$1_{\tau_B \le T} \le \frac{(S_T - \psi(B))^+}{B - \psi(B)} - \frac{1}{B - \psi(B)}(S_T - B)1_{\tau_B \le T}$$

with $\tau_B \equiv \inf\{t > 0 : S_t \ge B\}$. As explained before, this corresponds to
a semi-static super-replication. We buy a T-Vanilla call with payoff $(S_T - \psi(B))^+/(B - \psi(B))$ and delta-hedge the underlying at τ_B. This pathwise inequality was first obtained by Hobson in this pioneer paper [100]. He observed
first the following inequality which is valid for any $K < B$:

$$1_{\tau_B \le T} \le \frac{(S_T - K)^+}{B - K} - \frac{1}{B - K}(S_T - B)1_{\tau_B \le T}$$

Indeed,
(i): if $\tau_B > T$ (i.e., the stock does not hit the barrier before expiry), we have
$0 \le \frac{(S_T-K)^+}{B-K}$ which is trivially satisfied.
(ii): if $\tau_B \le T$, then the right-hand side becomes equal to (resp. greater than)
1 if $S_T > K$ (resp. $S_T \le K$).
Since the strike K is arbitrary, we can optimize over $K < B$. This reproduces
the optimal bound. Note that our stochastic control framework allows to
derive automatically this pathwise inequality by finding a super-solution to a
HJB equation. The additional pathwise inequality (4.29) gives

$$1_{M_T>B} \le \frac{(S_T - \psi(B))^+}{B - \psi(B)} - \int_0^T \frac{1_{\min(S_t,\psi(M_t) \ge \psi(B)}}{B - \psi(B)} dS_t$$

⬛

4.6.1.3 Forward-start Azéma–Yor

In this section, we provide a second application of our approach to the case
where the payoff $g(M_{t_2}^{1,2})$ depends on the forward running maximum between
t_1 and t_2: $M_t^{1,2} \equiv \max_{s \in [t_1,t]} S_s$. The model-free superhedging consistent
with t_1 and t_2 Vanillas can be explicitly computed:

THEOREM 4.5
Let $g(M)$ be some increasing C^1 function. Then,

$$\mathrm{MK}_2^c(\mathbb{P}^1, \mathbb{P}^2) = g(0) + \int C(t_2, \psi_2^*(m))g'(m)dm$$

with ψ_2^* the largest root of the equation

$$(\psi_2^*(y) - y)C(t_2, \psi_2^*(y)) = C(t_2, \psi_2^*(y)) - C(t_1, y) \qquad (4.35)$$

with $C(t_i, K) \equiv \mathbb{E}^{\mathbb{P}^i}[(S_{t_i} - K)^+]$, $i = 1, 2$.

PROOF (Sketch), Details can be found in [77]
(i) From the proof of Theorem 4.4 (see in particular equation (4.33)),

$$\sup_{\mathbb{P} \in \mathcal{M}^c} \mathbb{E}^{\mathbb{P}}[g(M_{t_2}^{1,2}) - \lambda_2(S_{t_2})|S_{t_1}, M_{t_1}^{1,2} \equiv S_{t_1}]] \leq g(S_{t_1}) - \lambda_2(S_{t_1})$$

$$\int_{\psi_2(S_{t_1})}^{S_{t_1}} (S_{t_1} - x)\lambda_2''(x)dx$$

where λ_2 given by $(\psi_2^*)'(x)\lambda_2''(\psi_2^*(x))(x - \psi_2^*(x)) = g'(x)$. From the dual representation of $\mathrm{MK}_2^c(\mathbb{P}^1, \mathbb{P}^2)$,

$$\mathrm{MK}_2^c(\mathbb{P}^1, \mathbb{P}^2) = \inf_{\lambda_2 \in L^1(\mathbb{P}^2)} \mathbb{E}^{\mathbb{P}^1}[\sup_{\mathbb{P} \in \mathcal{M}^c} \mathbb{E}^{\mathbb{P}}[g(M_{t_2}^{1,2}) - \lambda_2(S_{t_2})|S_{t_1}, M_{t_1}^{1,2} \equiv S_{t_1}]]$$

$$+ \mathbb{E}^{\mathbb{P}^2}[\lambda(S_{t_2})]$$

$$\leq \inf_{\lambda_2 \in L^1(\mathbb{P}^2)} \mathbb{E}^{\mathbb{P}^1}[g(S_{t_1}) - \lambda_2(S_{t_1}) + \int_{\psi_2(S_{t_1})}^{S_{t_1}} (S_{t_1} - x)\lambda_2''(x)dx]$$

$$+ \mathbb{E}^{\mathbb{P}^2}[\lambda(S_{t_2})]$$

Taking the infimum over ψ_2 as in Remark 4.5, we get our expression for ψ_2^* (4.35).
(ii): ψ_2^* induces a solution τ_2^* to SEP. We conclude the proof of the optimality of the upper bound derived above by arguing that

$$\mathrm{MK}_2^c(\mathbb{P}^1, \mathbb{P}^2) \geq \mathbb{E}^{\mathbb{P}^1}[\mathbb{E}[g(M_{\tau_2^*}^{1,2}) - \lambda_2(B_{\tau_2^*})|B_{t_1} = S_{t_1}, M_{t_1}^{1,2} = S_{t_1}]]$$

$$+ \mathbb{E}^{\mathbb{P}^2}[\lambda(S_{t_2})] = \mathbb{E}[g(M_{\tau_2^*}^{1,2})] \quad \text{as } B_{\tau_2^*} \sim \mathbb{P}^2$$

$\square$

This problem was solved in Hobson [100] in the case $g(x) = x$ using a different approach.

Example 4.2 Upper bound versus models
We have compared our optimal upper bound against the prices obtained using Dupire's local volatility model (in short LV) [68] and Bergomi's stochastic

Table 4.2: A digital option $g(m) = 100 \times 1_{m > \psi_\mu^{-1}(K)}$ on the maximum (maturity: 1 year, $\psi_\mu^{-1}(K = 118.44\% S_0) = S_0$). EuroStock (pricing date: 31/01/11). The optimal bound is $100 \times \frac{C^\mu(K)}{\psi_\mu^{-1}(K) - K}$.

LV	40.03
Bergomi($\sigma = 200\%$)	35.54
LV-Bergomi($\sigma = 200\%$)	41.25
LV-Bergomi($\sigma = 250\%$)	44.02
Upper	57.56

Table 4.3: Forward-start digital $g(M^{1,2}) = 100 \times 1_{M^{1,2} > K}$. $t_1 = 1$ year, $t_2 = 2$ years. EuroStock (pricing date: 31/01/11). The optimal bound is $100 \times C(t_2, \psi_2^*(K))$.

K/S_0	$\psi_2^*(K)/S_0$	LV	Upper
110%	88.74%	57.23	60.27
120%	102.31%	40.21	44.80
130%	115.67%	23.28	27.90
140%	127.78%	11.53	14.36
150%	140.15%	4.55	6.04
160%	151.57%	1.55	2.04
170%	161.53%	0.44	0.63

volatility model [24]. The LV model has been calibrated to EuroStock's implied volatility (31-Jan-2011). The Bergomi model has been calibrated to the variance-swap term structure: $T \mapsto -(2/T)\mathbb{E}^{\mathbb{P}^{\text{mkt}}}[\ln \frac{S_T}{S_0}]$. We have also included the prices produced by the local Bergomi's stochastic volatility model (in short LV-Bergomi), as introduced in [91], which has the property to be calibrated to the market implied volatilities. As expected, the prices are below our upper bound. Parameters for the Bergomi model: $\sigma = 2.0$ (or $\sigma = 2.5$, or $\sigma = 1.3$), $\theta = 22.65\%$, $k_1 = 4$, $k_2 = 0.125$, $\rho = 34.55\%$, $\rho_{SX} = -76.84\%$, $\rho_{SY} = -86.40\%$.

We have considered the payoff $g(m) = 100 \times 1_{m > \psi_\mu^{-1}(K)}$ (see Table 4.2), $g(M^{12}) = 100 \times 1_{M^{12} > K}$ (see Table 4.3) and $g(s, m) = 100 \times 1_{m \leq B}(s - K)^+$ (see Table 4.4).

□

Table 4.4: Lookback $g(s,m) = 100 \times 1_{m \leq B}(s - K)^+$ with $B > S_0$, $K < B$. EuroStock (pricing date: 31/01/11). The optimal bound is $100 \times \frac{(B-K)}{B-\psi_\mu(B)} C^\mu(\psi_\mu(B)) 1_{B > \psi_\mu^{-1}(K)} + C^\mu(K) 1_{\psi_\mu^{-1}(K) > B}$. Note that prices as given by the local volatility model are close to the optimal upper bound.

K/S_0	B/S_0	Upper	LV	Bergomi+LV($\sigma = 1.3$)
80%	106.8%	28.18	28.00	28.03
90%	106.8%	21.39	21.34	21.43
90%	109.5%	21.39	21.27	21.30
100%	106.8%	15.43	15.43	15.47
100%	109.5%	15.43	15.42	15.44
100%	112.5%	15.43	15.38	15.36
100%	115.8%	15.43	15.29	15.21
110%	112.5%	10.46	10.47	10.35
110%	115.8%	10.46	10.46	10.33
110%	119.4%	10.46	10.42	10.25
120%	123.4%	6.64	6.64	6.35
120%	128.0%	6.64	6.60	6.27
130%	133.7%	3.92	3.92	3.54
140%	140.4%	2.14	2.15	1.81

4.6.1.4 Multi-marginals

The key ingredient for the solution of $\mathrm{MK}_n^c(\mathbb{P}^1, \ldots, \mathbb{P}^n)$ turns out to be the following minimisation problem: for all $m \geq S_0$,

$$C(m) \equiv \min_{\zeta_1 \leq \cdots \leq \zeta_n \leq m} \sum_{i=1}^n \left(\frac{C(t_i, \zeta_i)}{m - \zeta_i} - \frac{C(t_i, \zeta_{i+1})}{m - \zeta_{i+1}} 1_{\{i < n\}} \right) \tag{4.36}$$

where $C(t_i, x) \equiv \mathbb{E}^{\mathbb{P}^i}[(S_{t_i} - x)^+]$. We report the following result:

THEOREM 4.6 see [94] and [121]
Let g be an increasing C^1 function and assume that $\mathbb{P}^1 \leq \ldots \leq \mathbb{P}^n$. Let $\zeta_1^*(m), \ldots, \zeta_n^*(m)$ be a solution to (4.36) for a fixed m. Then,

$$\mathrm{MK}_n^c(\mathbb{P}^1, \ldots, \mathbb{P}^n) \leq U \equiv g(S_0) \tag{4.37}$$

$$- \sum_{i=1}^n \int_{S_0}^\infty \left(\frac{C(t_i, \zeta_i^*(m))}{m - \zeta_i^*(m)} - \frac{C(t_i, \zeta_{i+1}^*(m))}{m - \zeta_{i+1}^*(m)} 1_{\{i < n\}} \right) dg(m)$$

Moreover, there exist $\lambda_i \in L^1(\mathbb{P}^i)$, and a trading strategy $H \in \mathcal{H}$, explicitly given in [94], such that $\mathrm{MK}_n^c(\mathbb{P}^1, \ldots, \mathbb{P}^n) = g(S_0) + \sum_{i=1}^n \mathbb{E}^{\mathbb{P}^i}[\lambda_i(S_{t_i})]$ and for *all $\omega \in \Omega$,*

$$U + \sum_{i=1}^{n} \lambda_i(S_{t_i}) - \sum_{i=1}^{n} \mathbb{E}^{\mathbb{P}^i}[\lambda_i(S_{t_i})] + \sum_{i=1}^{n} H_{t_{i-1}}(S_{t_i} - S_{t_{i-1}}) \geq g(M_T)$$

Assume further that $\mathbb{P}^1, \ldots, \mathbb{P}^n$ satisfy Assumption A in [121]. Then, equality holds in (4.37).

The limit $n \to \infty$ is studied in [110].

PROOF (Sketch), for $n = 2$
(i): From Proposition 4.1, we have

$$\mathrm{MK}_2^c(\mathbb{P}^1, \mathbb{P}^2) = \inf_{(\lambda_i(\cdot))_{i=1,\ldots,2}} u_0(S_0, S_0) + \sum_{i=1}^{2} \mathbb{E}^{\mathbb{P}^i}[\lambda_i(S_{t_i})]$$

where $u_0(s, m) \equiv \sup_{\mathbb{P} \in \mathcal{M}^c} \mathbb{E}[g(M_{t_2}) - \sum_{i=1}^2 \lambda_i(S_{t_i})]$. The dynamic programming principle implies that u_0 can be obtained through

$$u_2(s, m) = g(m)$$
$$u_1(s, m) = \sup_{\mathbb{P} \in \mathcal{M}^c} \mathbb{E}[u_2(S_{t_2}, M_{t_2}) - \lambda_2(S_{t_2})|S_{t_1} = s, M_{t_1} = s]$$
$$u_0(s, m) = \sup_{\mathbb{P} \in \mathcal{M}^c} \mathbb{E}[u_1(S_{t_1}, M_{t_1}) - \lambda_1(S_{t_1})|S_0 = s, M_0 = s]$$

From the proof of Theorem 4.4 (see equation (4.33) in the case of a payoff $g(s, m)$), we have that u_k can be bounded by U_k defined iteratively by

$$U_k(s, m) = U_{k+1}(s, m) - \lambda_{k+1}(s)$$
$$+ \int_{\psi_{k+1}(m)}^{\max(s, \psi_{k+1}(m))} (s - x)\left(\lambda_{k+1}''(x) - \partial_1^2 U_{k+1}(x, m)\right) dx \quad (4.38)$$

where $\psi_{k+1}(m) < m$ is the maximal solution of the ODE

$$\psi_{k+1}'\left(\lambda_{k+1}''(\psi_{k+1}) - \partial_1^2 U_{k+1}(\psi_{k+1}, m)\right)(m - \psi_{k+1}) = \partial_2 U_{k+1}(\psi_{k+1}, m) +$$
$$(m - \psi_{k+1})\partial_{12} U_{k+1}(\psi_{k+1}, m)$$

By solving these recursive equations, we get an upper bound

$$\mathrm{MK}_2^c(\mathbb{P}^1, \mathbb{P}^2) \leq g(S_0) + \inf_{\psi_1(y), \psi_2(y) < y} \int_{S_0}^{\infty} \frac{C(t_2, \psi_2(y))}{y - \psi_2(y)} dg(y) \quad (4.39)$$
$$- \int_{S_0}^{\infty} \left(\frac{C(t_1, \psi_2(y))}{y - \psi_2(y)} - \frac{C(t_1, \psi_1(y))}{y - \psi_1(y)}\right) 1_{\psi_1(y) < \psi_2(y)} dg(y)$$

(ii): Taking the pointwise minimisation over ψ_i, we obtain

$$\psi_{\mathbb{P}^1}^{-1}(y) = \frac{\mathbb{E}^{\mathbb{P}^1}[S_{t_1} 1_{S_{t_1} > y}]}{\mathbb{E}^{\mathbb{P}^1}[1_{S_{t_1} > y}]} \tag{4.40}$$

$$\psi_{\mathbb{P}^2}^{-1}(y) = \frac{\mathbb{E}^{\mathbb{P}^2}[S_{t_2} 1_{S_{t_2} > y}]}{\mathbb{E}^{\mathbb{P}^2}[1_{S_{t_2} > y}]}, \quad \text{for } \psi_{\mathbb{P}^1}^{-1}(y) < \psi_{\mathbb{P}^2}^{-1}(y) \tag{4.41}$$

$$\psi_{\mathbb{P}^2}^{-1}(y) = \frac{\mathbb{E}^{\mathbb{P}^2}[S_{t_2} 1_{S_{t_2} > y}] - \mathbb{E}^{\mathbb{P}^1}[S_{t_1} 1_{S_{t_1} > y}]}{\mathbb{E}^{\mathbb{P}^2}[1_{S_{t_2} > y}] - \mathbb{E}^{\mathbb{P}^1}[1_{S_{t_1} > y}]} \quad \text{otherwise} \tag{4.42}$$

This defines stopping time $\tau_1 < \tau_2$ such that $B_{\tau_i} \sim \mathbb{P}^i$ and for which the above bound is optimal.

Note that Proposition 4.1 is not needed as in Section 4.5 as one needs only to check that our pathwise super-replication strategy is valid and then use the weak duality result (4.9). $\qquad \square$

Note that the derivation of the multi-marginal SEP requires some technical assumptions on $\mathbb{P}^1, \dots, \mathbb{P}^n$ (see Assumption A in [121]). The multi-marginal SEP is completely solved in the case of the Root embedding [52] following this approach.

Example 4.3 $g(m) = 1_{m \geq B}$
From (4.39), $(\psi_{\mathbb{P}^1} \equiv \psi_{\mathbb{P}^1}(B), \psi_{\mathbb{P}^2} \equiv \psi_{\mathbb{P}^2}(B))$

$$\mathrm{MK}_2^c(\mathbb{P}^1, \mathbb{P}^2) = \frac{C(t_2, \psi_{\mathbb{P}^2})}{B - \psi_{\mathbb{P}^2}} + \left(\frac{C(t_1, \psi_{\mathbb{P}^2})}{B - \psi_{\mathbb{P}^2}} - \frac{C(t_1, \psi_{\mathbb{P}^1})}{B - \psi_{\mathbb{P}^1}} \right) 1_{\psi_{\mathbb{P}^1} < \psi_{\mathbb{P}^2}}$$

This corresponds to the pathwise super-replication obtained from (4.38): for all $\mathbb{P} \in \mathcal{M}^c$,

$$1_{M_{t_2} \geq B} \leq \frac{(S_{t_2} - \psi_{\mathbb{P}^2})^+}{B - \psi_{\mathbb{P}^2}} + \left(\frac{(S_{t_1} - \psi_{\mathbb{P}^2})^+}{B - \psi_{\mathbb{P}^2}} - \frac{(S_{t_1} - \psi_{\mathbb{P}^1})^+}{B - \psi_{\mathbb{P}^1}} \right) 1_{\psi_{\mathbb{P}^1} < \psi_{\mathbb{P}^2}}$$

$$+ 1_{M_{t_1} \geq B} \frac{B - S_{t_1}}{B - \psi_{\mathbb{P}^1}} + 1_{M_{t_1} \geq B} 1_{S_{t_1} \geq \psi_{\mathbb{P}^2}} \frac{S_{t_1} - S_{t_2}}{B - \psi_{\mathbb{P}^2}}$$

$$+ 1_{M_{t_1} \leq B} 1_{M_{t_2} \geq B} \frac{B - S_{t_2}}{B - \psi_{\mathbb{P}^2}}, \quad \mathbb{P} - \text{a.s.}$$

This inequality is easily checked by considering separately each case. This reproduces the result in [32]. The first line consists in holding t_1 and t_2 Vanilla calls. The term $1_{M_{t_1} \geq B} \frac{B - S_{t_1}}{B - \psi_{\mathbb{P}^1}}$ (resp. $1_{M_{t_1} \leq B} 1_{M_{t_2} \geq B} \frac{B - S_{t_2}}{B - \psi_{\mathbb{P}^2}}$) corresponds to performing a delta-hedging when $\tau_B \leq t_1$ (resp. $\tau_B \in (t_1, t_2]$). The term $1_{M_{t_1} \geq B} 1_{S_{t_1} \geq \psi_{\mathbb{P}^2}} \frac{S_{t_1} - S_{t_2}}{B - \psi_{\mathbb{P}^2}}$ corresponds to a delta-hedging at t_1. $\qquad \square$

4.6.2 Root's solution

4.6.2.1 Barrier

The Root solution consists in finding a solution of SEP (i.e., $B_{\tau_R} \sim \mu$) where $\tau_R = \inf\{t > 0 \;:\; (t, B_t) \in R\}$ with $R = \{(t, x) \mid t \geq r(x)\}$ for some function $r : \mathbb{R} \to [0, \infty]$. The function r can be explicitly characterized in terms of a variational PDE or a nonlinear equation as found in [53, 69, 80, 81]

PROPOSITION 4.8 [53, 69, 80, 81]
r is characterized as the unique solution of the equation:

$$\int (y - x)^+ \mu(dy) - (S_0 - x)^+ = \frac{1}{2} g(r(x), S_0 - x) \qquad (4.43)$$

$$- \frac{1}{2} \int_{\mathbb{R}} \mu(dy) 1_{r(x) \geq r(y)} g(r(x) - r(y), y - x) dy$$

with $g(t, x) = \sqrt{\frac{2t}{\pi}} e^{-\frac{x^2}{2t}} - |x| + x(N\left(\frac{x}{\sqrt{2t}}\right) - 1)$ *where* $N(\cdot)$ *is the normal cumulative function.*

This equation can be seen as an early premium formula for a perpetual American option.

PROOF Set $u(t, x) \equiv \mathbb{E}[(B_{\tau_R \wedge t} - x)^+]$. From Tanaka's lemma, we have

$$d(B_{\tau_R \wedge t} - x)^+ = 1_{B_t \geq x} 1_{t < \tau_R} dB_t + \frac{1}{2} dL_t^x 1_{t < \tau_R}$$

where L_t^x is the local time at x (formally $dL_t^x = \delta(B_t - x)dt$). By taking the expectation, we get

$$\partial_t u(t, x) = \frac{1}{2} \mathbb{E}[\delta(B_{t \wedge \tau_R} - x) 1_{t < \tau_R}] = \frac{1}{2} \partial_x^2 u(t, x) - \frac{1}{2} \mathbb{E}[\delta(B_{\tau_R} - x) 1_{t \geq \tau_R}]$$

Finally as $r(B_{\tau_R}) = \tau_R$, we have

$$\mathbb{E}[\delta(B_{\tau_R} - x) 1_{t \geq \tau_R}] = \mathbb{E}[\delta(B_{\tau_R} - x) 1_{t \geq r(B_{\tau_R})}]$$
$$= 1_{t \geq r(x)} \mathbb{E}[\delta(B_{\tau_R} - x)] = 1_{t \geq r(x)} \mu(x) \text{ as } B_{\tau_R} \sim \mu$$

and

$$\partial_t u(t, x) = \frac{1}{2} \partial_x^2 u(t, x) - \frac{1}{2} \mu(x) 1_{t \geq r(x)}, \quad u(0, x) = (S_0 - x)^+ \quad (4.44)$$

This is nothing else than a semi-linear PDE associated to a variational PDE - obstacle problem [53, 69]:

$$\max\left(u^\mu(x) - u(t, x), \partial_t u(t, x) - \frac{1}{2} \partial_x^2 u\right) = 0$$

with $u^\mu(x) \equiv \mathbb{E}^\mu[(S_T - x)^+]$. The solution of PDE (4.44) is

$$u(t, x) = \int_\mathbb{R} (S_0 - y)^+ p(t|y - x)dy - \frac{1}{2} \int_\mathbb{R} \mu(dy) \int_0^t dsp(t - s|y - x)1_{s \geq r(y)}$$

$$(4.45)$$

where $p(t|x) \equiv \frac{1}{\sqrt{2\pi t}} e^{-\frac{x^2}{2t}}$ the fundamental solution of the heat kernel. At the boundary, we have $u(r(x), x) = u^\mu(x)$ (and the smooth-fit condition $\partial_x u(r(x), x) = \partial_x u^\mu(x)$). By taking $t = r(x)$, this implies that

$$u^\mu(x) = \int_\mathbb{R} (S_0 - y)^+ p(r(x)|y - x)dy$$

$$- \frac{1}{2} \int_\mathbb{R} \mu(dy) \int_0^{r(x)} dsp(r(x) - s|y - x)1_{s \geq r(y)}$$

By setting $g(t, x) \equiv \int_0^t p(s|x)ds$ and Fubini's theorem, we see that

$$\int_\mathbb{R} \mu(dy) \int_0^{r(x)} dsp(r(x) - s|y - x)1_{s \geq r(y)} =$$

$$\int_\mathbb{R} \mu(dy)g(r(x) - r(y), y - x)1_{r(x) \geq r(y)}$$

Furthermore,

$$\int_\mathbb{R} (S_0 - y)^+ \left(p(r(x)|y - x) - \delta(y - x) \right) dy =$$

$$\int_\mathbb{R} (S_0 - y)^+ dy \int_0^{r(x)} \partial_t p(t|y - x)dt =$$

$$\frac{1}{2} \int_\mathbb{R} (S_0 - y)^+ \int_0^{r(x)} \partial_y^2 p(t|y - x)dydt \overset{\text{IBP}}{=} \frac{1}{2}g(r(x), S_0 - x)$$

In the second line, we have used that p is solution of the heat kernel, i.e., $\partial_t p(t|y - x) = \frac{1}{2}\partial_y^2 p(t|y - x)$. In the last line, we have applied an integration by parts formula. Finally, this gives (4.43). □

In fact, this representation can be simplified into

PROPOSITION 4.9
r is characterized as the unique solution of the equation:

$$p(r(x)|x - S_0) = \int_\mathbb{R} 1_{r(x) \geq r(y)} p(r(x) - r(y)|x - y)\mu(dy)$$

where $p(t|x) \equiv \frac{1}{\sqrt{2\pi t}} e^{-\frac{x^2}{2t}}$ is the fundamental solution of the heat kernel.

PROOF   The two smooth-fit conditions

$$u(r(x), x) = u^\mu(x)$$
$$\partial_x u(r(x), x) = \partial_x u^\mu(x)$$

satisfied by the solution (4.45) imply that $\partial_t u(r(x), x) = 0$. By differentiating PDE (4.44) with respect to t, we obtain for $\Theta(t, x) \equiv \partial_t u(t, x)$:

$$\partial_t \Theta(t, x) = \frac{1}{2}\partial_x^2 \Theta(t, x) - \frac{1}{2}\mu(x)\delta(t - r(x)), \quad \Theta(0, x) = \frac{1}{2}\delta(S_0 - x)$$

From Feynman–Kac formula, we get

$$\Theta(t, x) = \frac{1}{2}p(t|x - S_0) - \frac{1}{2}\int_{\mathbb{R}} 1_{t \geq r(y)}p(t - r(y)|x - y)\mu(dy) \quad (4.46)$$

Using that $\Theta(r(x), x) = 0$ as derived previously, we get that the barrier $r(\cdot)$ satisfies our new integral representation. $\square$

4.6.2.2   Optimality

THEOREM 4.7 Optimality of the Root solution [131, 53]
Let g be some C^1 concave function. Define

$$U(t, s) \equiv \int_0^t \mathbb{E}_{u,s}[g'(\tau_R)]du - Z(s), \quad Z''(s) \equiv 2\mathbb{E}_{0,s}[g'(\tau_R)] \quad (4.47)$$
$$\lambda^*(s) \equiv g(r(s)) - U(r(s), s) \quad (4.48)$$

(i): For all $\mathbb{P} \in \mathcal{M}^c$,

$$U(0, S_0) + \int_0^T \partial_s U(t, S_s)dS_s \geq g(\langle S \rangle_T) - \lambda^*(S_T)$$

(ii):

$$\mathrm{MK}_1^c(\mu) = U(0, S_0) + \mathbb{E}^\mu[\lambda^*(S_T)] = \mathbb{E}^{\mathbb{P}^*}[g(\langle S \rangle_T)]$$

and the bound is attained for $\mathbb{P}^*$ defined as the distribution of $S_t = B_{\tau_R \wedge \frac{t}{T-t}}$.

Note that Z is defined up to an irrelevant affine function. The optimal lower bound is computed in [54] and is attained by the Rost solution to SEP [132].

PROOF
(i): Following the proof outlined in Section 4.5, an upper bound is obtained if we can find a supersolution to the variational PDE:

$$\max(g(t) - \lambda(s) - u(t, s), \mathcal{L}u) = 0 \quad (4.49)$$

with $\mathcal{L} \equiv \frac{1}{2}\partial_s^2 + \partial_t$. As an ansatz, we are looking for a solution of the form:

$$\frac{1}{2}\partial_s^2 U(t,s) + \partial_t U(t,s) = 0, \quad \forall\, (t,s) \in \mathbb{R}_+^2 \tag{4.50}$$

$$U(r(s),s) = g(r(s)) - \lambda(s) \; : \; \text{(instantaneous stopping)}$$

that we complete by the additional requirement (i.e., the delta is smooth at the boundary)

$$\partial_s U(r(s),s) = -\lambda'(s) \; : \; \text{(smooth fit)}$$

Then, we can check that $\Theta(t,s) \equiv \partial_t U(t,s)$ satisfies the same PDE (4.50) with the boundary $\Theta(r(s),s) = g'(r(s))$ (see a similar result in the proof of Proposition 4.9). Dynkin's formula implies that

$$\Theta(t,s) = \mathbb{E}_{t,s}[g'(\tau_{\mathrm{R}})]$$

By integration, we get U as given by (4.47). The instantaneous stopping condition corresponds to (4.48). The super-replication strategy is then obtained using Itô's lemma under the concavity assumption for g. In particular $U(t,s) \geq g(t) - \lambda^*(s)$ (i.e., $U(t,s) - U(r(s),s) \geq g(t) - g(r(s))$) as

$$\int_{r(s)}^{t} du \mathbb{E}_{u,s}\Big[\int_u^{\tau_{\mathrm{R}}} g''(v) dv\Big] \geq 0$$

(ii): The optimality is proved from the weak duality (4.9) and by checking that $U(0,S_0) + \mathbb{E}^\mu[\lambda^*(S_T)] = \mathbb{E}[g(\tau_{\mathrm{R}})]$. Note that by construction $S_T = B_{\tau_{\mathrm{R}}} \sim \mu$.
$\square$

In the next solution, we explain how to derive (analytical) bounds for more general payoffs $g(S_T, \langle \ln S \rangle_T)$. As our main concern is about applications in finance, we consider here the quadratic variation of $\ln S_T$.

4.6.2.3 (Non-optimal) analytical bounds

Let's take a domain $\mathcal{D} \subset \mathbb{R}_+^2$ such as $(S_0, V_0 = 0) \in \mathcal{D}$, $\mathbb{E}[\tau_\mathcal{D}] < \infty$ with $\tau_\mathcal{D} \equiv \inf\{t > 0 \; : \; (t, B_t^{\mathrm{geo}}) \notin \mathcal{D}\}$. B^{geo} denotes a geometric Brownian motion. We define for $g \in C^{2,1}$

$$\underline{\lambda}''(s) \equiv \frac{2}{s^2} \sup_{v \notin \mathcal{D}} \left(\frac{s^2}{2}\partial_s^2 g(s,v) + \partial_v g(s,v) \right), \quad \forall\, s \notin \mathcal{D} \tag{4.51}$$

$$\overline{\lambda}''(s) \equiv \frac{2}{s^2} \inf_{v \in \mathcal{D}} \left(\frac{s^2}{2}\partial_s^2 g(s,v) + \partial_v g(s,v) \right), \quad \forall\, s \in \mathcal{D} \tag{4.52}$$

Here $s \in \mathcal{D}$ (resp. $v \in \mathcal{D}$) means that there exists $v \in \mathbb{R}_+$ (resp. $s \in \mathbb{R}_+$) such that $(s,v) \in \mathcal{D}$.

Assumption 7

$$\underline{\lambda}''(s) \leq \overline{\lambda}''(s), \quad \forall \, s \text{ such that } \exists \, (v, v') \, : \, (s, v) \in \mathcal{D} \text{ and } (s, v') \notin \mathcal{D}$$

THEOREM 4.8 Upper bound of $g(x, v)$
Under Assumption 7, an upper bound for $\mathrm{MK}_1^c(\mu)$ is given by

$$\mathrm{MK}_1^c(\mu) \leq \mathbb{E}[g(B_{\tau_\mathcal{D}}^{\mathrm{geo}}, \tau_\mathcal{D})] + \int_0^\infty dx \underline{\lambda}''(x) \left(C^\mu(x) - \mathbb{E}[(B_{\tau_\mathcal{D}}^{\mathrm{geo}} - x)^+] \right)^+ 1_{x \notin \mathcal{D}}$$

$$- \int_0^\infty dx \overline{\lambda}''(x) \left(\mathbb{E}[(B_{\tau_\mathcal{D}}^{\mathrm{geo}} - x)^+] - C^\mu(x) \right)^+ 1_{x \in \mathcal{D}}$$

with $C^\mu(x) = \mathbb{E}^\mu[(S_T - x)^+]$.

Similarly, a lower bound can be derived by replacing g by $-g$. The term $\mathbb{E}[(B_{\tau_\mathcal{D}}^{\mathrm{geo}} - x)^+] - C^\mu(x)$ represents the difference between a timer-like call option and the market price of a call option with the same strike x.

PROOF   Let $\mathcal{D}$ be a domain in (s, v) and $\tau_\mathcal{D}$ the corresponding exit time with $\mathbb{E}[\tau_\mathcal{D}] < \infty$. Following the proof outlined in Section 4.5, an upper bound is obtained if we can find a supersolution to the variational PDE:

$$\max \left(g(s, v) - \lambda(s) - u(s, v), \mathcal{L}u \right) = 0 \tag{4.53}$$

with $\mathcal{L} \equiv \frac{s^2}{2} \partial_s^2 + \partial_v$. As an ansatz, we take

$$U^\lambda(s, v) \equiv \mathbb{E}_{s,v}[g^\lambda(B_{\tau_\mathcal{D}}^{\mathrm{geo}}, \tau_\mathcal{D})], \quad \forall \, (s, v) \in \mathcal{D}$$
$$= g^\lambda(s, v), \quad \forall \, (s, v) \notin \mathcal{D} \tag{4.54}$$

with $g^\lambda \equiv g - \lambda$. We need to show that (4.54) verifies the obstacle problem (4.53):
(i): $\forall \, (s, v) \in \mathcal{D}, \quad \mathcal{L}U^\lambda(s, v) = 0$ by construction.
(ii): Applying $\mathcal{L}$ on the function $\psi(s, v) \equiv \mathbb{E}_{s,v}[g^\lambda(B_{\tau_\mathcal{D}}^{\mathrm{geo}}, \tau_\mathcal{D})] - g^\lambda(s, v)$ defined on $\mathcal{D}$, we get

$$\mathcal{L}\psi(s, v) + \mathcal{L}g^\lambda(s, v) = 0, \quad \forall \, (s, v) \in \mathcal{D}, \quad \psi|_{\partial \mathcal{D}} = 0$$

which from Dynkin's formula can be represented as

$$\psi(s, v) = \mathbb{E}_{s,v}[\int_0^{\tau_\mathcal{D}} \mathcal{L}g^\lambda(B_u^{\mathrm{geo}}, u) du]$$

We deduce that (4.53) is satisfied if

$$\mathcal{L}g^\lambda(s, v) \geq 0, \quad \forall \, (s, v) \in \mathcal{D}$$
$$\mathcal{L}g^\lambda(s, v) \leq 0, \quad \forall \, (s, v) \notin \mathcal{D}$$

We obtain inequalities obeyed by the function $\lambda''(\cdot)$:

$$\underline{\lambda}''(s) \le \lambda''(s), \quad \forall\, s \notin \mathcal{D} \tag{4.55}$$

$$\lambda''(s) \le \overline{\lambda}''(s), \quad \forall\, s \in \mathcal{D} \tag{4.56}$$

The supersolution is (by assuming that $(S_0, V_0 = 0) \in \mathcal{D}$)

$$U^\lambda(S_0, V_0 = 0) = \mathbb{E}[g^\lambda(B^{\text{geo}}_{\tau_\mathcal{D}}, \tau_\mathcal{D})]$$

$$= \mathbb{E}[g(B^{\text{geo}}_{\tau_\mathcal{D}}, \tau_\mathcal{D})] - \int_0^\infty dx\, \lambda''(x)\mathbb{E}[(B^{\text{geo}}_{\tau_\mathcal{D}} - x)^+]$$

and for $\lambda(\cdot)$ satisfying inequalities (4.55), (4.56):

$$\text{MK}_1^c(\mu) \le \mathbb{E}[g(B^{\text{geo}}_{\tau_\mathcal{D}}, \tau_\mathcal{D})] + \int_0^\infty dx\, \lambda''(x)\left(C^\mu(x) - \mathbb{E}[(B^{\text{geo}}_{\tau_\mathcal{D}} - x)^+]\right)$$

Taking the infimum over the function $\lambda(\cdot)$, we derive our upper bound. $\quad\square$

Example 4.4 $g(s, v) = -\psi(s)(v - K)^+$, $\psi(s) > 0$ and convex

We take $\mathcal{D} = \mathbb{R}_+ \times [0, K]$. Therefore, we get $(V_{\tau_\mathcal{D}} - K)^+ = 0$ and $\mathbb{E}[(S_{\tau_\mathcal{D}} - x)^+] = C_{\text{BS}}(\sigma^2_{\text{BS}}T = K, x)$ with C_{BS} the Black–Scholes price of a call option with strike x and variance $\sigma^2_{\text{BS}}T = K$ (i.e., timer call, see Section 3.3.1). Finally, we have $\underline{\lambda}(s) = -\frac{2\psi(s)}{s^2}$, $\overline{\lambda}(s) = 0\ \forall\, s \in \mathbb{R}_+$. Assumption 7 is satisfied and we obtain the *lower* bound for $-g$:

$$2\int_0^\infty \frac{\psi(x)}{x^2}\left(C^\mu(x) - C^{\text{BS}}(\sigma^2_{\text{BS}}T = K, x)\right)^+ dx$$

This formula reproduces Dupire's result [69] for the specific case $\psi(x) = 1$. For $g(s, v) = (s - K)^+ v$, we get the lower bound

$$\sup_{\bar{V}\in\mathbb{R}_+} \bar{V}C^\mu(K) + 2\int_K^\infty \frac{(x - K)}{x^2}\left(C^\mu(x) - C^{\text{BS}}(\sigma^2_{\text{BS}}T = \bar{V}, x)\right) dx$$

$\quad\square$

Example 4.5
Upper bound: $g(s, v) = (v - K)^+$
In order to reproduce Carr–Lee result [45], we take $\mathcal{D} = [\underline{s}, \overline{s}] \times \mathbb{R}_+$ with $\underline{s} < S_0 < \overline{s}$ and consider the modified payoff $\tilde{g}(s, v) = (v - K)^+ + \Lambda(s)$, the European payoff Λ being left unspecified for the moment. We have

$$\underline{\lambda}''(s) \equiv \Lambda''(s) + \frac{2}{s^2}, \quad \forall\, s \notin [\underline{s}, \overline{s}]$$

$$\overline{\lambda}''(s) \equiv \Lambda''(s), \quad \forall\, s \in [\underline{s}, \overline{s}]$$

With the choice $\Lambda(s) = 2\left(\ln s - \ln \overline{s} - \frac{\ln \frac{\overline{s}}{\underline{s}}}{\overline{s} - \underline{s}}(s - \overline{s})\right)^-$, we get

$$\underline{\lambda}(s) = 0, \quad \forall s \notin [\underline{s}, \overline{s}], \quad \overline{\lambda}(s) = 0, \quad \forall s \in [\underline{s}, \overline{s}]$$

for which we deduce an upper bound on $g(s, v)$:

$$\inf_{\underline{s} < \overline{s}} \left\{ \mathbb{E}[(\tau_{\mathcal{D}} - K)^+] - \mathbb{E}^\mu[\Lambda(S_T)] \right\}$$

▯

Example 4.6

Lower/Upper bound: $g(s, v) = \frac{s}{S_0} v^\alpha$

We take $\mathcal{D} = \mathbb{R}_+ \times [0, \bar{V}]$ and we have $\underline{\lambda}(s) = \overline{\lambda}(s) = \frac{2\alpha}{sS_0}\overline{V}^{\alpha - 1}$ for $\alpha \in [0, 1]$. Applying our formula, we get the upper bound

$$\bar{V}^\alpha + \alpha \bar{V}^{\alpha - 1}\left(2\mathbb{E}^\mu[\frac{S_T}{S_0}\ln\frac{S_T}{S_0}] - \bar{V}\right)$$

Taking the infimum over $\bar{V} \in \mathbb{R}_+$, we have the upper bound

$$\left(2\mathbb{E}^\mu[\frac{S_T}{S_0}\ln\frac{S_T}{S_0}]\right)^\alpha, \quad \alpha \in [0, 1]$$

Proceeding similarly, we can prove that this expression is a lower bound for $\alpha \notin (0, 1)$ (see Figure 4.9).

▯

We have compared our lower/upper bounds (see Figures 4.3, 4.4) against the fair values obtained using Dupire's local volatility [68] (in short LV) and Bergomi's two-factor stochastic volatility model [24]. The LV model (resp. Bergomi's model) has been calibrated to EuroStock's implied volatility (2-Feb-2010) (resp. variance-swap term structure $T \mapsto -(2/T)\mathbb{E}^{\mathbb{P}^{\text{mkt}}}[\ln \frac{S_T}{S_0}]$). We observed that the fair values as given by Bergomi's model can be below our lower bound - indicating that the spot/volatility correlations ρ_{SX} and ρ_{SY} have been chosen improperly and the implied volatility as produced by the Bergomi's model is not inline with the market (see Figures 4.3, 4.4, 4.5). We have added the fair values produced by local Bergomi's stochastic volatility model, as introduced in [91], which has the property to be calibrated to the market implied volatility. As expected, the fair values are above our lower bound in this case.

4.6.3   Perkins solution

Set $M_t \equiv \max_{0 < s < t} B_t$ and $m_t \equiv \min_{0 < s < t} B_t$. The Perkins solution to SEP is

$$\tau_{\text{Perkins}} = \inf\{t > 0 \ : \ B_t \notin (\gamma_+(M_t), \gamma_-(m_t))\}$$

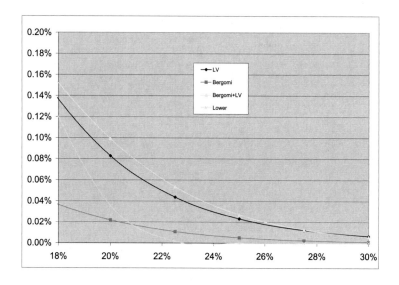

Figure 4.3: Lower bounds for option $g(s, v) = (\frac{s}{S_0} - 1)^+ (v - K)^+$ as a function of K. Parameters for the Bergomi model: $\theta = 22.65\%$, $k_1 = 4$, $k_2 = 0.125$, $\rho = 34.55\%$, $\rho_{SX} = -76.84\%$, $\rho_{SY} = -86.40\%$, $\sigma = 2.0$. As Bergomi's model is not calibrated to Vanilla options here, it can produce prices below our lower bounds.

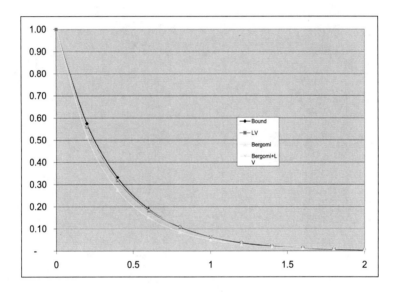

Figure 4.4: Lower/upper bound for options: $g(s,v) = \frac{s}{S_0}v^\alpha$ as a function of $\alpha \in [0,2]$. Parameters for the Bergomi model: $\theta = 22.65\%$, $k_1 = 4$, $k_2 = 0.125$, $\rho = 34.55\%$, $\rho_{SX} = -76.84\%$, $\rho_{SY} = -86.40\%$, $\sigma = 2.0$. As Bergomi's model is not calibrated to Vanilla options here, it can produce prices below (resp. upper) our lower (resp. upper) bounds.

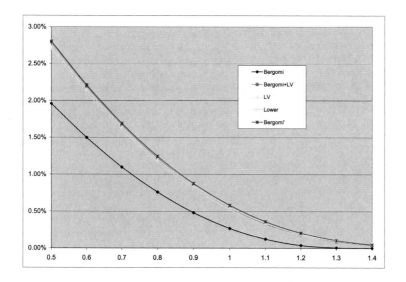

Figure 4.5: Lower for option: $g(s, v) = (s - K)^+ v$ as a function of K. Parameters for the Bergomi model: $\theta = 22.65\%$, $k_1 = 4$, $k_2 = 0.125$, $\rho = 34.55\%$, $\rho_{SX} = -76.84\%$, $\rho_{SY} = -86.40\%$, $\sigma = 2.0$. As Bergomi's model is not calibrated to Vanilla options here, it produces prices below our lower bounds. Bergomi' denotes the Bergomi model calibrated to call options $T \mapsto \mathbb{E}^\mu[(S_T - K)^+]$ instead of the variance swap term structure. In this case, the price is greater than our lower bound.

with $\gamma_+(j) = \mathrm{argmax}_{x<S_0}\frac{C^\mu(j)-P^\mu(x)}{j-x}$ and $\gamma_-(i) = \mathrm{argmin}_{x>S_0}\frac{P^\mu(i)-C^\mu(x)}{x-i}$.
Here $C^\mu(x) \equiv \mathbb{E}^\mu[(S_T - x)^+]$ and $P^\mu(x) \equiv \mathbb{E}^\mu[(x - S_T)^+]$. The Perkins embedding has the property that it simultaneously minimises the law of the maximum M and maximises the law of the minimum m:

THEOREM 4.9 Optimality of the Perkins solution [127]
Let τ be a solution to SEP. Then for all increasing function g,

$$\mathbb{E}[g(M_\tau)] \geq \mathbb{E}[g(M_{\tau_{\mathrm{Perkins}}})]$$
$$\mathbb{E}[g(m_\tau)] \leq \mathbb{E}[g(m_{\tau_{\mathrm{Perkins}}})]$$

Example 4.7 One-touch digital, $g(M) = 1_{M>B}$
The Perkins construction is based on the following pathwise sub-replication [31] for $K < B$:

$$1_{\tau_B \leq T} \geq 1_{S_T \geq B} + \frac{(S_T - B)^+}{B - K} - \frac{(K - S_T)^+}{B - K} + \frac{B - S_T}{B - K}1_{\tau_B \leq T}$$

Taking the expectation, we obtain for all $\mathbb{P} \in \mathcal{M}^c(\mu)$

$$\mathbb{E}^\mathbb{P}[1_{\tau_B \leq T}] \geq -C^\mu(B)' + \frac{C^\mu(B) - P^\mu(K)}{B - K}$$

Since this is valid for all $K < B$, we can maximize over K and find

$$\mathbb{E}^\mathbb{P}[1_{\tau_B \leq T}] \geq -C^\mu(B)' + \sup_{K<B}\frac{C^\mu(B) - P^\mu(K)}{B - K}$$

This bound is attained by the martingale measure defined by $S_t = B_{\frac{t}{T-t}\wedge\tau_{\mathrm{Perkins}}}$.
See also [31] for further discussion of barrier options and [51] for applications to two-sided barriers. ◻

4.6.4 Vallois' solution

Given $\phi_+ : \mathbb{R}_+ \to [S_0, \infty)$ and $\phi_- : \mathbb{R}_+ \to [0, S_0]$ satisfying Assumption 8 below, define

$$\gamma(l) \equiv \frac{1}{2}\int_0^l \left(\frac{1}{\phi_+(m) - S_0} - \frac{1}{\phi_-(m) - S_0}\right)dm \quad \text{for all } l > 0.$$

Assumption 8
(i) ϕ_+ (resp. ϕ_-) is continuous and strictly increasing (resp. decreasing).
(ii) $\gamma(0+) = 0$ and $\gamma(+\infty) = +\infty$.

Under Assumption 8, we define the static strategy $\lambda : \mathbb{R}_+ \to \mathbb{R}$ via (ϕ_+, ϕ_-) by

$$\lambda'(S_0 \pm x) - \lambda'(S_0\pm) \equiv \int_0^{\psi_\pm(S_0\pm x)} \frac{dy}{\phi_\pm(y) - S_0} e^{\gamma(y)} \int_y^{+\infty} e^{-\gamma(m)} g''(dm)$$

(4.57)

$$\pm\lambda'(S_0\pm) \equiv g'(0) + \int_0^{+\infty} e^{-\gamma(m)} g''(dm)$$

(4.58)

$$\lambda(S_0) \equiv g(0).$$

(4.59)

where $\psi_\pm$ denote the inverse of $\phi_\pm$ and $g''(dm)$ is the second derivative in the sense of distribution of g. $g : \mathbb{R}_+ \to \mathbb{R}$ is assumed to be a convex function with uniformly bounded derivative denoted by g'. Notice that the restriction of λ on (S_0, ∞) and $[0, S_0)$ is convex. Following the same game as in Section 4.5, we can compute explicitly $\mathrm{MK}_1^c(\mu)$.

PROPOSITION 4.10 Robust super-replication
Under Assumption 8, the following inequality holds

$$\int_0^T H_u dS_u + \lambda(S_T) \geq g(L_T), \quad \mathbb{P} \text{ - a.s. for all } \mathbb{P} \in \mathcal{M}^c, \qquad (4.60)$$

where the function λ is defined by (4.57), (4.58), (4.59) and the process H is given either by

$$H_u \equiv -\lambda'(\phi_+(L_u))\mathbf{1}_{\{S_u > S_0\}} - \lambda'(\phi_-(L_u))\mathbf{1}_{\{S_u \leq S_0\}}, \qquad (4.61)$$

or by

$$H_u \equiv -\lambda'(\max(S_u, \phi_+(L_u)))\mathbf{1}_{\{S_u > S_0\}} - \lambda'(\min(S_u, \phi_-(L_u)))\mathbf{1}_{\{S_u \leq S_0\}}$$

(4.62)

PROOF Following closely Section 4.5, the proof consists in finding a supersolution to the variational PDE:

$$\max\left(g(l) - \lambda(s) - U(s, l), \frac{1}{2}\partial_s^2 U + \delta(s - S_0)\partial_l U(s, l)\right) = 0$$

See [48] for details and [50] for an alternative approach. The reader should not care about the mathematical justification of the above PDE, in particular the term $\delta(s - S_0)\partial_l U(s, l)$. Once a supersolution will be found, the proof will consist in using Tanaka's formula to prove that U is a well-defined supersolution. As an ansatz, we look for a solution satisfying:

$$\frac{1}{2}\partial_s^2 U + \delta(s - S_0)\partial_l U(s, l) = 0, \quad s \in [\phi_-(l), \phi_+(l)] \qquad (4.63)$$

$$U(\phi_\pm(l), l) = g(l) - \lambda(\phi_\pm(l))$$

$$\partial_s U(\phi_\pm(l), l) = -\lambda'(\phi_\pm(l))$$

Outside the domain $(\phi_-(l), \phi_+(l))$, we should have $-\lambda''(s)/2 + \delta(s - S_0)g'(l) = -\lambda''(s)/2 \leq 0$. Therefore λ should be convex. The appearance of the delta function $\delta(s - S_0)$ indicates that the function U should involve an at-the-money call payoff $(s - S_0)^+$. We take a solution of the form:

$$U(s, l) = a(l)(s - S_0)^+ + b(l)(s - S_0) + c(l)$$

Plugging U into equations (4.63), we deduce that

$$a(l) = -2c'(l)$$
$$b(l) = -\lambda'(\phi_-(l))$$
$$c(l) = g(l) - \lambda(\phi_-(l)) + \lambda'(\phi_-(l))(\phi_-(l) - S_0)$$
$$\lambda'(\phi_+) - \lambda'(\phi_-) = 2g'(l) + 2\lambda''(\phi_-)\phi_-'(\phi_- - S_0)$$
$$\lambda(\phi_+) - \lambda(\phi_-) - \lambda'(\phi_-)(\phi_+ - \phi_-) = 2(\phi_+ - S_0)(g'(l) + \lambda''(\phi_-)\phi_-'(\phi_- - S_0))$$
$$(4.64)$$

By integrating the two last equations, we obtain our equations (4.57), (4.58), (4.59) for λ. In particular, λ is convex if and only if g is convex. Finally, we have for all $s \in [\phi_-(l), \phi_+(l)]$:

$$U(s, l) = -(\lambda'(\phi_+) - \lambda'(\phi_-))(s - S_0)^+ - \lambda'(\phi_-)(s - \phi_-) - \lambda(\phi_-) + g(l)$$

and $U(s, l) = g(l) - \lambda(s)$ outside. By applying Tanaka's lemma on U, we get inequality (4.60) with H equal to (4.62). Similarly, if we take for all $s \in \mathbb{R}_+$:

$$U(s, l) = -(\lambda'(\phi_+) - \lambda'(\phi_-))(s - S_0)^+ - \lambda'(\phi_-)(s - \phi_-) - \lambda(\phi_-) + g(l)$$

we get inequality (4.60) with H equal to (4.61).   ꠶

Let $(B_t)_{t \geq 0}$ be a Brownian motion and denote by $(L_t^B)_{t \geq 0}$ its local time at zero. It follows by Vallois [137] that there exists a pair (ϕ_+^μ, ϕ_-^μ) such that $B_{\tau_{\text{Vallois}}} \sim \mu$, where

$$\tau_{\text{Vallois}} \equiv \inf \left\{ t > 0 : B_t \notin (\phi_-^\mu(L_t^B), \phi_+^\mu(L_t^B)) \right\}. \qquad (4.65)$$

In addition, if μ has no atoms, then (ϕ_+^μ, ϕ_-^μ) satisfies Assumption 8. If μ admits a positive density $\mu(x)$ with respect to the Lebesgue measure, one can show from Lemma 3.2 (take $\lambda(s) = \delta(s - K)$ and $\mathbb{E}[\delta(B_{\text{Vallois}} - K)] = \mu(K)$ - see a similar proof in 4.6.1.1) that

$$\phi_\pm^{\mu\,\prime}(l) = \frac{1 - F(\phi_+^\mu(l)) + F(\phi_-^\mu(l))}{2(\phi_\pm^\mu(l) - S_0)\mu(\phi_\pm^\mu(l))}, \qquad \text{for all } l > 0,$$

$$\phi_\pm^\mu(0) = S_0.$$

where $F(x) \equiv \mu((0, x])$ is the cumulative distribution of μ. If we assume further that μ is symmetric, then $\phi_\pm^\mu(x) = S_0 \pm \phi^\mu(x)$ and

$$\psi^\mu(x) = \int_0^x \frac{y\mu(y)}{1 - F(y)} dy,$$

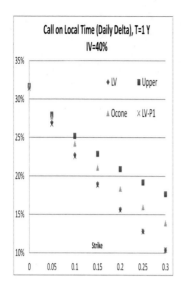
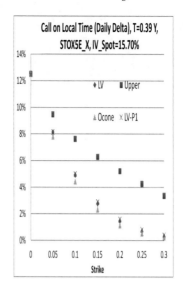

Figure 4.6: Optimal upper bound for a call on local time $(L_T - K)^+$ as a function of the strike K. Left: μ is a log-normal distribution with $\sigma_{\mathrm{BS}} = 0.4$ and $T = 1$ year. Right: μ is implied from market prices of call options on EuroStock with $T = 0.39$ year. Parameters for the BergomiLV(=LV-P1) model: $\sigma = 174\%$, $\theta = 0.24$, $k_1 = 5.35$, $k_2 = 0.28$, $\rho = 0$, $\rho_{\mathrm{SX}} = \rho_{\mathrm{SY}} = -0.7$.

where ψ^μ denotes the inverse of ϕ^μ.

THEOREM 4.10 Optimality
Let g be a convex function with uniformly bounded derivative. Then,

$$\mathrm{MK}_1^c(\mu) = \mathbb{E}^\mu[\lambda^\mu] = \mathbb{E}^{\mathbb{P}^*}[g(L_T)]$$

where λ^μ is constructed from (ϕ_+^μ, ϕ_-^μ). $\mathbb{P}^*$ is defined as the distribution of $S_t = B_{\tau_{\mathrm{Vallois}}} \wedge \frac{t}{T-t}$. In addition,the inequality (4.60) is attained for $\mathbb{P}^*$:

$$\int_0^T H_u^\mu dS_u + \lambda^\mu(S_T) = g(L_T), \ \mathbb{P}^* - a.s., \tag{4.66}$$

where the process H^μ is given by (4.61) (or (4.62)) with λ^μ and (ϕ_+^μ, ϕ_-^μ).

PROOF   Use the weak duality (4.9).   ☐

The 2-marginals case is considered in [48].

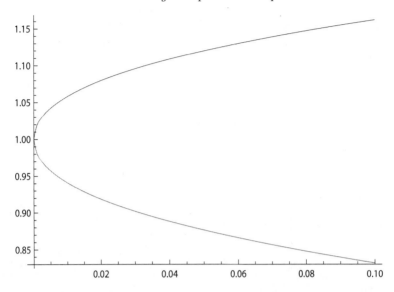

Figure 4.7: Vallois' embedding ϕ_+^μ/S_0 (top curve) and ϕ_-^μ/S_0 where μ is implied from market prices of call options on EuroStock with $T = 0.39$ year.

In Figure 4.6, we have computed $\mathrm{MK}_1^c(\mu)$ for $F_T = (L_T - K)^+$ (for different strikes K). We compared these bounds against prices produced by the Dupire Local volatility model (denoted LV), the local Bergomi model (denoted LV-P1) which has the property to be calibrated to the implied volatility surface. The Vallois embedding (ϕ_+^μ, ϕ_-^μ) is reported in Figure 4.7.

We have also reported the prices given by Ocone martingales with T-marginal μ (in this case the price is model-independent, see Section 3.4.3). Note that the T-marginal μ implied from market prices of call options on EuroStock is not symmetric in log-moneyness and therefore the computation with Ocone martingales is not justified. It is however reported for completeness.

4.7   Matching marginals through SEP

In Section 4.2, we have listed examples of martingale measures (i.e., arbitrage-free models) fitting marginals (i.e., implied volatilities). Other constructions can be obtained through solutions of SEP. More precisely, let τ be a solution of SEP(μ). Then define the martingale process S_t by $S_t \equiv B_{\tau \wedge \frac{t}{T-t}}^{\mathrm{geo}}$. Trivially, we have $S_T = B_\tau^{\mathrm{geo}} \sim \mu$. Taking for τ the Azéma–Yor, Vallois, Perkins, Root, Rost solutions to SEP, we know also that the corresponding measure $\mathbb{P}^*$

satisfies an optimality result for a certain class of payoffs. Here, based on the Azéma–Yor and Vallois solutions, we build examples in $\mathcal{M}^{\text{cadlag}}((\mathbb{P}^t)_{t\in(0,T]})$. We characterize their infinitesimal generators which correspond to local Lévy models as described previously.

4.7.1 Through Azéma–Yor

For a given family of marginals $(\mathbb{P}^t)_{0\leq t\leq T}$, we define the family of barycenter functions:

$$\psi^{-1}(t,x) \equiv \frac{\mathbb{E}^{\mathbb{P}^t}[S_t 1_{S_t\geq x}]}{\mathbb{E}^{\mathbb{P}^t}[1_{S_t\geq x}]}$$

We set $\tau_t \equiv \inf\{u > 0 \; : \; M_u \geq \psi^{-1}(t, B_u)\}$.

Assumption 9
(i) *for all* $0 \leq s \leq t \leq T$, $\psi(s,\cdot) \leq \psi(t,\cdot)$, *i.e.*, $\tau_s \leq \tau_t$.
(ii) *the map* $(t,x) \mapsto \psi(t,x)$ *admits first-order partial derivatives.*

Note that condition (1) is not always satisfied if $\mathbb{P}^s \leq \mathbb{P}^t$.

THEOREM 4.11
Under Assumptions 9 for ψ, $(B_{\tau_t})_{0\leq t\leq T}$ *is an inhomogeneous Markov martingale process with generator* $\mathcal{L}f(x)$ *given by*

$$\frac{\partial_t\psi(t,x)}{\partial_x\psi(t,x)}\left(\frac{1}{\psi(t,x)-x}\frac{\int_x^\infty \mathbb{P}^t(dy)\,(f(y)-f(x))\,dy}{\int_x^\infty \mathbb{P}^t(dy)dy} - f'(x)\right)$$

for all smooth bounded functions $f : \mathbb{R} \to \mathbb{R}$. *In particular, the process* $(B_{\tau_t})_{0\leq t\leq T}$ *is a jump Markov martingale, which corresponds to an example of local Lévy's model as introduced in Carr, Geman, Madan and Yor [42].*

PROOF see [113] for an alternative proof using excursion theory
Let us calculate the generator of B_{τ_t} by a PDE approach. For any test function f, we evaluate

$$\mathcal{A}_t(f)(s) \equiv \frac{d}{du}|_{u=t}\mathbb{E}[f(B_{\tau_u})|B_{\tau_t} = s]$$

We have (see Lemma 3.1)

$$U(u,s,m) \equiv \mathbb{E}[f(B_{\tau_u})|B_0 = s, M_0 = m]$$
$$= \alpha_u(m)(s - \psi_u(m)) + f(\psi_u(m))$$

where

$$\alpha_u(m) = \int_m^\infty e^{\int_{m'}^m \frac{d\psi_u(n)}{n-\psi_u(n)}} \frac{f(\phi_u)d\psi_u(m')}{m' - \psi_u(m')}$$

with $\psi_u(m) \equiv \psi(u, m)$. Note that $\mathbb{E}[f(B_{\tau_u})|B_{\tau_t} = s] = U(u, s, \psi^{-1}(t, s))$. Finally, by differentiating $U(u, s, \psi^{-1}(t, s))$ with respect to u and by taking $u = t$, we get our result. ⬚

4.7.2   Through Vallois

In this section, for the sake of simplicity, we assume that μ_t is symmetric and absolutely continuous with respect to the Lebesgue measure for every $t \in (0, T]$. Denote

$$\psi(t, x) = \int_0^x \frac{y\mu_t(y)}{1 - \mu_t((-\infty, y])} dy,$$

THEOREM 4.12
Under Assumptions 9 for ψ, $(B_{\tau_t})_{0 \leq t \leq T}$ is an inhomogeneous Markov martingale process with generator $\mathcal{L}f$ given by

$$\frac{\partial_t \psi_t(|x|)}{\partial_x \psi_t(|x|)} \left(\frac{e^{\gamma_t(\psi_t(|x|))}}{2|x|} \int_{|x|}^{+\infty} (f(y) + f(-y) - 2f(x)) \, de^{-\gamma_t(\psi_t(y))} \right.$$

$$\left. - \text{sgn}(x)f'(x) \right)$$

for all smooth bounded functions $f : \mathbb{R} \to \mathbb{R}$.

Details can be found in [48].

4.7.3   Optimality in $\mathcal{M}^c((\mathbb{P}^t)_{t \in (0,T]})$

Assumption 9 is usually satisfied by marginals inferred from market implied volatilities, smoothly interpolated. Under this assumption, we can show that the robust (sub)super-replication price depends only on the terminal marginal. In particular intermediate Vanilla options are not needed.

COROLLARY 4.3 [113]
Under Assumptions 9 for ψ and for $g(m)$ an increasing function depending on the running maximum,

$$\text{MK}_\infty^c((\mathbb{P}^t)_{t \in (0,T]}) = \text{MK}_1^c(\mathbb{P}^T)$$

Similar result for $g(l)$ a convex function depending on the local time.

PROOF
(i): $\text{MK}_\infty^c((\mathbb{P}^t)_{t \in (0,T]}) \leq \text{MK}_1^c(\mathbb{P}^T)$ as the super-replication strategy derived for MK_1^c applies to MK_∞^c.

(ii): As $B_{\tau_t} \sim \mathbb{P}^t$ for all $t \in [0, T]$ and τ_t is increasing in t

$$\mathrm{MK}_\infty^c((\mathbb{P}^t)_{t \in (0,T]}) \geq \mathbb{E}[g(M_{\tau_T})] = \mathrm{MK}_1^c$$

The last equality is obtained as τ_T coincides with the Azéma–Yor solution. ⬜

4.8 Martingale inequalities

Mass transport provides a power tool to study some functional inequalities with geometric content. Let us mention the optimal Sobolev inequality, entropy-entropy production inequalities and transport inequalities (see Chapter 9 in [139]). In this section, we explain how MOT can provide similar results for martingale inequalities. We illustrate this on Doob's inequality and Burkholder–Davis–Gundy inequality for $p = 1$. As a by-product, we obtain pathwise inequalities. Alternative approaches are presented in [1, 21, 15, 19]. For a payoff ξ, let us define

DEFINITION 4.4

$$\mathrm{MK}_1^c(p, \mu) \equiv \inf \left\{ Y_0 : \exists\, \lambda \in \mathbb{R} \text{ and } H \in \mathcal{H}, \right.$$
$$\left. \overline{Y}_T^{H,\lambda} \geq \xi, \mathbb{P} - a.s. \text{ for all } \mathbb{P} \in \mathcal{M}^c \right\},$$

where $\overline{Y}^{H,\lambda}$ *denotes the portfolio value of a self-financing strategy with continuous trading* H *in the underlying, and static trading in a p-payoff* S_T^p *with market price* $\mathbb{E}^\mu[S_T^p]$ *:*

$$\overline{Y}_T^{H,\lambda} \equiv Y_0 + \int_0^T H_s dS_s + \lambda\, (S_T^p - \mathbb{E}^\mu[S_T^p]) \tag{4.67}$$

4.8.1 Doob's inequality revisited

Let S_t be a càdlàg positive martingale starting at S_0 at $t = 0$. Doob's inequality states that

$$\mathbb{E}[M_T^p] \leq S_0^p + \left(\frac{p}{p-1}\right)^p \mathbb{E}[S_T^p], \quad \forall\, p > 1$$

The derivation of this inequality (with the optimal constant $p/(p-1)$) can be framed into the computation of a robust upper bound for the payoff M_T^p within the class of arbitrage-free models calibrated to T-power options with payoff S_T^p:

THEOREM 4.13

(i) *(Optimal bound):*

$$\mathrm{MK}_1^c(p,\mu) = -\frac{p}{p-1}S_0^p + \left(\frac{p}{p-1}\right)^p \mathbb{E}^\mu[S_T^p]$$

(ii) *(Pathwise inequality): For all* $\mathbb{P} \in \mathcal{M}^c(\mu)$:

$$\mathrm{MK}_1^c(p,\mu) - \frac{p^2}{p-1}\int_0^T M_s^{p-1}dS_s + \left(\frac{p}{p-1}\right)^p (S_T^p - \mathbb{E}^\mu[S_T^p]) \geq M_T^p, \quad \mathbb{P}-a.s.$$

A similar inequality was provided in [1] using an alternative approach.

PROOF   Following our discussion on Azéma–Yor solution (see Section 4.6.1.2), we have

$$\mathrm{MK}_1^c(p,\mu) \equiv \inf_{\lambda \in \mathbb{R}} u^\lambda(S_0, S_0) + \lambda \mathbb{E}^\mu[S_T^p]$$

where $u^\lambda(s,m) \equiv \sup_{\tau \in \mathcal{T}} \mathbb{E}[M_\tau^p - \lambda B_\tau^p | B_0 = s, M_0 = m]$. u^λ is a viscosity solution of the variational inequality:

$$\max\left(m^p - \lambda s^p - u^\lambda(s,m), \partial_s^2 u^\lambda(s,m)\right) = 0, \quad s \leq m$$

with the Neumann condition $\partial_m u^\lambda(m,m) = 0$. Using a similar ansatz for the optimal stopping time as in Azéma–Yor solution, we derive the solution (4.33) and ψ^λ satisfies (4.30) with $\lambda(x) \equiv \lambda x^p$. Solutions are $\psi^\lambda(m) = c\,m$ with $\lambda(p-1)c^{p-1}(1-c) = 1$ and $c \in (0,1)$. This implies that

$$\mathrm{MK}_1^c(p,\mu) \leq -\frac{S_0^p}{(1-c)(p-1)} + \frac{1}{(p-1)c^{p-1}(1-c)}\mathbb{E}[S_T^p]$$

Taking the infimum over $c \in (0,1)$, we obtain

$$\mathrm{MK}_1^c(p,\mu) \leq -\frac{p}{p-1}S_0^p + \left(\frac{p}{p-1}\right)^p \mathbb{E}[S_T^p]$$

The supremum was reached for $c^* = (p-1)/p$ and $\lambda^* = p^p/(p-1)^p$. Finally we check that our stopping time $\tau^* = \inf\{t > 0 : B_t \leq c^* B_t^*\}$ is optimal. From our stochastic control approach, we obtain the pathwise inequality: for all $\mathbb{P} \in \mathcal{M}^c(\mu)$,

$$\mathrm{MK}_1^c(p,\mu) - \int_0^T \lambda'(\psi^\lambda(M_s))dS_s + \lambda^* (S_T^p - \mathbb{E}[S_T^p]) \geq M_T^p$$

We conclude as $\lambda'(\psi^\lambda(M_s)) = \frac{p^2}{p-1}M_s^{p-1}$.   ▯

4.8.2 Burkholder–Davis–Gundy inequality

Another classical result in stochastic analysis is the Burkholder–Davis–Gundy (in short BDG) inequality which states that for any $p \geq 0$ there exists universal constants $c_1(p)$ and $c_2(p)$ such that for all local martingales S with $S_0 \equiv 0$ and for all T, the following inequality holds

$$c_1(p)\mathbb{E}[V_T^{\frac{p}{2}}] \leq \mathbb{E}[M_T^p] \leq c_2(p)\mathbb{E}[V_T^{\frac{p}{2}}]$$

Here $V_t \equiv \langle S \rangle_t$ denotes the quadratic variation and $M_t \equiv \max_{0 < s < t} S_t$ the running maximum. For continuous local martingales, this statement holds for all $p > 0$.

As stated in Peskir, Shiryaev [128], "the question of finding the best possible values for $c_1(p)$ and $c_2(p)$ appears to be of interest. Its emphasis is not so much on having these values but more on finding a method of proof which can deliver them." In the case where $\mathbb{E}[M_T^p]$ is replaced by $\mathbb{E}[S_T^p]$, the optimal constants can be computed by the method of time change as described in [128, 55], Section 10.

For the BDG inequalities, only partial results are known: for $p = 2$, the best constants are $c_1(2) = 1$ and $c_2(2) = 4$ (See [128] Section 20). One is deduced from Doob's inequality and the other from the stopping time $\tau = \inf\{t > 0 \ : \ |B_t| = 1\}$ with B a one-dimensional Brownian motion. Quite recently, for $p = 1$, Burkholder [33] obtains $c_1(1) = 1/\sqrt{3}$ and Osekowski [123] $c_2(1) \simeq 1.30693$. Peccati and Yor [126] shows that if we consider the subclass of continuous martingale, the constant $1/\sqrt{3}$ is no longer optimal and could be replaced by $1/\sqrt{2}$. In this section, we reproduce the optimal constant for $p = 1$ (for a refined BDG version) and provide also an (optimal) pathwise version of this inequality. The case $p > 1$ is still in progress. Some (non-optimal) pathwise inequalities have been found in [21].

THEOREM 4.14 Optimal BDG $p = 1$, see also [123] Theorem 4.1

(i) *For all* $\mathbb{P} \in \mathcal{M}^c$,

$$\mathbb{E}^{\mathbb{P}}[M_T] \leq S_0 + c_2(1)\mathbb{E}^{\mathbb{P}}[\sqrt{V_T}] \tag{4.68}$$

with $c_2(1) \simeq 1.30693$ *is the zero of*

$$M(-c_2(1)) = 0 \tag{4.69}$$

Here $M(x) \equiv M(-\frac{1}{2}, \frac{1}{2}, \frac{x^2}{2})$ *where* $M(a,b,c)$ *is Kummer's confluent hypergeometric function. This bound corresponds to the pathwise inequality*

$$S_0 + c_2(1)\mathbb{E}^{\mathbb{P}}[\sqrt{V_T}] + c_2(1)\left(\sqrt{V_T} - \mathbb{E}^{\mathbb{P}}[\sqrt{V_T}]\right)$$
$$+ \int_0^T \left(-\frac{M'\left(\frac{S_t - M_t}{\sqrt{V_t}}\right)}{M'(-c_2(1))} + 1\right) dS_t \geq M_T, \quad \mathbb{P} - a.s. \ \text{for all } \mathbb{P} \in \mathcal{M}^c \tag{4.70}$$

(ii) *The bound is attained by* $\mathbb{P}^*$ *defined as the distribution of* $S_t = B_{\tau^* \wedge \frac{t}{T-t}}$
with $\tau^* = \inf\{t > 0 \ : \ B_t \le M_t - c_2(1)\sqrt{t}\}$.

Note that if we take the expectation of the inequality (4.70), we obtain (4.68).
The above pathwise inequality (4.70) has a nice interpretation in mathematical finance. The BDG inequality (for $p = 1$) corresponds to the computation
of a model-independent bound for a lookback payoff M_T with maturity T consistent with the (market) price of (normal) volatility swap with payoff $\sqrt{V_T}$.
This bound can be super-replicated by a strategy involving at $t = 0$ a cash
amount $S_0 + c_2(1)\mathbb{E}[\sqrt{V_T}]$, a static hedge of the payoff $\sqrt{V_T}$ with weight $c_2(1)$
and finally a delta-hedging $H_t \equiv \left(-\dfrac{M'\left(\frac{S_t - M_t}{\sqrt{V_t}}\right)}{M'(-c_2(1))} + 1 \right)$.

PROOF   Our derivation follows closely our section on Azéma–Yor solution
and Doob's inequality. Here we consider the general case $p \ge 1$ and specialized
to $p = 1$ later. We have

$$\mathbb{E}[M_T^p] = \inf_{\lambda \in \mathbb{R}} \mathbb{E}[M_T^p - \lambda V_T^{\frac{p}{2}}] + \lambda \mathbb{E}[V_T^{\frac{p}{2}}] \le \inf_{\lambda \in \mathbb{R}} \sup_{\tau} \mathbb{E}[M_\tau^p - \lambda V_\tau^{\frac{p}{2}}] + \lambda \mathbb{E}[V_T^{\frac{p}{2}}]$$

We set below $u^\lambda(s, m, v) \equiv \sup_\tau \mathbb{E}[M_\tau^p - \lambda V_\tau^{\frac{p}{2}} | S_0 = s, M_0 = m, V_0 = v]$ and

$$\mathbb{E}[M_T^p] \le \inf_{\lambda \in \mathbb{R}} u^\lambda(S_0, S_0, 0) + \lambda \mathbb{E}[V_T^{\frac{p}{2}}]$$

$u^\lambda(s, m, v)$ solves the variational PDE

$$\max\left(m^p - \lambda v^{\frac{p}{2}} - u^\lambda, \frac{1}{2}\partial_s^2 u^\lambda + \partial_v u^\lambda \right) = 0, \quad s \le m$$

subject to the Neumann condition $\partial_m u(m, m, v) = 0$. As stated in Peskir
and Shiryaev [128], section 20, this non-linear problem is inherently three-
dimensional and quite difficult to solve. In fact this PDE can be converted into
a two-dimensional PDE, quite similar to the variational PDE (4.24) appearing
in Azéma–Yor solution. We simplify this equation with the change of variables
$u^\lambda(s, m, v) = v^{\frac{p}{2}} U^\lambda \left(x \equiv \frac{s}{\sqrt{v}}, y \equiv \frac{m}{\sqrt{v}}, z \equiv \frac{1}{\sqrt{v}} \right)$. It is clear that the function
$u^\lambda(s, m, v)$ should scale as α^p when $s \to \alpha s$, $m \to \alpha m$ and $v \to \alpha^2 v$ from its
definition. This scaling symmetry implies that the function U^λ is independent
of z and therefore solves a *two*-dimensional PDE: For all $y \in \mathbb{R}_+$,

$$\max\left(y^p - \lambda - U^\lambda(x, y), \partial_x^2 U^\lambda + p U^\lambda - x\partial_x U^\lambda - y\partial_y U^\lambda \right) = 0, \quad x \le y$$

subject to the Neumann condition $\partial_y U^\lambda(y, y) = 0$. The derivation of the
optimal BDG inequality is then achieved by guessing a solution of the above

equation of the form

$$\partial_x^2 U^\lambda + pU^\lambda - x\partial_x U^\lambda - y\partial_y U^\lambda = 0, \quad \partial_y U^\lambda(y,y) = 0$$
$$U^\lambda(\psi_\lambda(y), y) = y^p - \lambda, \quad \partial_x U^\lambda(\psi_\lambda(y), y) = 0 \quad (4.71)$$

where the optimal stopping time is

$$\tau_{\psi_\lambda} = \inf\left\{t > 0 \ : \ S_t \le \sqrt{V_t}\psi_\lambda\left(\frac{M_t}{\sqrt{V_t}}\right)\right\}$$

We look for a solution depending on the variables x, y only through the variables x and $x - y$. For $p = 1$, the most general solution is

$$U^\lambda = aM(x - y) + b(x - y) + AM(x) + Bx$$

depending on 4 constants a, b, A and B. The Neumann condition gives that $b = 0$. Furthermore $\lim_{v \to 0} U^\lambda(S_0, S_0, v) < \infty$ if and only if $A = 0$. As we should have $aM(x - y) + Bx > y - \lambda$ for all $x - y \le 0$, we must have $B = 1$. Conditions (4.71) give

$$aM(\psi_\lambda(y) - y) + \psi_\lambda(y) = y - \lambda$$
$$aM'(\psi_\lambda(y) - y) + 1 = 0$$

We get

$$U^\lambda(x, y) = -\frac{M(x - y)}{M'(\psi_\lambda(y) - y)} + x$$

with ψ_λ such that

$$\lambda = \frac{M(\psi_\lambda(y) - y)}{M'(\psi_\lambda(y) - y)} - (\psi_\lambda(y) - y)$$

Taking the infimum of λ over $\psi_\lambda(y) - y$, we get $\lambda^* = -(\psi_\lambda(y) - y)$ with $M(-\lambda^*) = 0$. This gives Equation (4.69). The super-replication property could be justified from a classical verification argument using Itô's formula. The analysis of the case $p > 1$ deserves further study. □

4.8.3 Inequalities on local time

Similar inequalities on local time can be derived. They gave some estimation of the moments of the local time by the moments of the underlying process. We set

$$c_1 = \mathbf{1}_{\{a_1 \ge a_2\}}\left(\frac{a_1^{1-p}}{1 - (p-1)a_1} - \frac{a_2^{1-p}}{1 - (p-1)a_2}\right) \text{ and } c_2 = \frac{a_2^{1-p}}{1 - (p-1)a_2}.$$

PROPOSITION 4.11
With the notations above, if we assume further that $c_1 \geq 0$, then the following pathwise inequality holds: for all $\mathbb{P} \in \mathcal{M}^c$, $\mathbb{P}$-a.s.,

$$L_{t_2}^p \leq c_1|S_{t_1}|^p + c_2|S_{t_2}|^p - \int_0^{t_2} p\Big(\mathbf{1}_{\{s \leq t_1\}}c_1 a_1^{p-1} + c_2 a_2^{p-1}\Big)|L_s|^{p-1}\mathrm{sgn}(S_s)\, dS_s$$

Further, one has

$$\mathbb{E}^{\mathbb{P}}[L_{t_2}^p] \leq c_1\mathbb{E}^{\mathbb{P}}[|S_{t_1}|^p] + c_2\mathbb{E}^{\mathbb{P}}[|S_{t_2}|^p].$$

In addition, the infimum of the r.h.s. above over all pairs (a_1, a_2) is attained for $a_1 = a_2 = \frac{1}{p}$, which yields

$$\mathbb{E}^{\mathbb{P}}[L_T^p] \quad \leq \quad p^p\mathbb{E}^{\mathbb{P}}[|S_T|^p].$$

The above inequality originates from [138].

The proof can be found in [48].

4.9   Randomized SEP

In Section 3.4, we have seen that by restricting the class $\mathcal{M}^c(\mu)$ to the subset $\mathcal{M}^c(0, \mu)$, corresponding to Ocone's martingales with fixed T-marginal μ, some path-dependent payoffs have a unique price characterized by a dynamic replication strategy with static positions in T-Vanillas. Switching to $\mathcal{M}^c(\mu)$, this model-independence breaks down and arbitrage-free prices range in the interval $[\underline{\mathrm{MK}}_1^c(\mu), \overline{\mathrm{MK}}_1^c(\mu)]$ where $\overline{\mathrm{MK}}_1^c(\mu)$ (resp. $\underline{\mathrm{MK}}_1^c(\mu)$) is the robust seller (resp. buyer) super-replication price. Here we define a new class $\mathcal{M}(\rho, \mu)$ parameterized by a constant ρ, interpreted as a spot/volatility correlation, such that $\mathcal{M}^c(\rho = 0, \mu) = \mathcal{M}^c(0, \mu)$ and $\mathcal{M}^c(\rho = 1, \mu) = \mathcal{M}^c(\mu)$. We consider buyer and seller's super-replication prices $(\underline{\mathrm{MK}}_1^c(\mu, \rho), \overline{\mathrm{MK}}_1^c(\mu, \rho))$ within this class, which interpolate from the super-replication price (for $\rho = 1$) and the replication price (for $\rho = 0$). Finding the optimal measure for $\overline{\mathrm{MK}}_1^c(\mu, \rho)$ will consist in finding a solution to a randomized SEP.

Class $\mathcal{M}^c(\rho)$ - *Conditional independence*: Fixed $\rho \in [-1, 1]$. We define the probability measure $\mathbb{P}^a$ where

$$dS_t = a_t(\rho dB_t + \sqrt{1 - \rho^2}dB_t^\perp)$$

and where $B^\perp$ is a Brownian motion orthogonal to B, a and $B^\perp$ are also *independent*. We denote by $\mathcal{M}^c(\rho)$ the class of such martingale probability measures and $\mathcal{M}^c(\rho, \mu) \equiv \mathcal{M}^c(\rho) \cap \mathcal{M}^c(\mu)$. Note that $\mathcal{M}^c(\rho = 0) = \mathcal{M}^c(0)$.

The class $\mathcal{M}^c(\rho)$ includes stochastic volatility models where the volatility dynamics is given by

$$da_t = b(a_t) + \sigma(a_t)dB_t$$

Here b (resp. σ) is the drift (resp. volatility-of-volatility), eventually multi-dimensional. Note that local stochastic volatility models do not belong to $\mathcal{M}^c(\rho, \mu)$. We also consider $\mathcal{M}^c(\rho, q)$ the subset $\mathcal{M}^c(\rho)$:

$$\mathcal{M}^c(\rho, q) \equiv \{\mathbb{P} \in \mathcal{M}^c(\rho) \ : \ \mathbb{E}^{\mathbb{P}}[S_T^{p\pm}] = \mathbb{E}^{\mu}[S_T^{p\pm}]\}$$

where p_+ and p_- denote the two solutions of $p^2(1 - \rho^2) \equiv q$ with $q \in \mathbb{R}_+$. We have the chain rule

$$\mathcal{M}^c(\rho, \mu) \subset \mathcal{M}^c(\rho, q) \subset \mathcal{M}^c(\rho) \subset \mathcal{M}^c.$$

4.9.1 Robust pricing with partial information

We define the primal problem for a payoff depending on the quadratic variation $\langle S \rangle_T$:

$$\mathrm{MK}(\rho, \mu) \equiv \sup_{\mathbb{P} \in \mathcal{M}^c(\rho, \mu)} \mathbb{E}[g(\langle S \rangle_T)]$$

Below, we consider a specific payoff, $g_q(\langle S \rangle_T) \equiv e^{\frac{q}{2}\langle S \rangle_T}$ and define the relaxed primal problem:

$$\mathrm{MK}_q(\rho) \equiv \sup_{\mathbb{P} \in \mathcal{M}^c(\rho, q)} \mathbb{E}[e^{\frac{q}{2}\langle S \rangle_T}]$$

Note that, under appropriate conditions, a general payoff $g(\langle S \rangle_T)$ can be decomposed over an exponential basis:

$$g(v) = \int_0^\infty dq g^*(q) e^{\frac{q}{2}v}$$

This implies that if $g^* > 0$,

$$\mathrm{MK}(\rho, \mu) \leq \int_0^\infty dq g^*(q) \mathrm{MK}_q(\rho) \tag{4.72}$$

For use below, we remind you of the well-known mixing formula, corresponding to a filtering with respect to the filtration generated by $B^\perp$:

LEMMA 4.1 Mixing formula

For all $\mathbb{P} \in \mathcal{M}^c(\rho)$,

$$\mathbb{E}_t^{\mathbb{P}}[e^{pS_T}] = e^{p(S_t - \bar{S}_t) - \frac{q}{2}\langle S \rangle_t} \mathbb{E}_t[e^{p\bar{S}_T + \frac{q}{2}\langle S \rangle_T}] \tag{4.73}$$

where the process $\bar{S}_t$ satisfies $d\bar{S}_t := \rho a_t dB_t$, $\quad \bar{S}_0 = S_0$.
In particular, for $t = 0$:

$$\mathbb{E}^{\mathbb{P}}[e^{pS_T}] = \mathbb{E}[e^{p\bar{S}_T + \frac{q}{2}\langle S \rangle_T}] \tag{4.74}$$

PROOF   By conditioning on the filtration generated by B_t and a_t, $S_T - S_0 - \int_0^T a_s dB_s = \sqrt{1-\rho^2}\int_0^T a_s dB_s$ is a Gaussian variable G with zero mean which can be integrated out explicitly. Then, we use the identity $\mathbb{E}[e^{sG}] = e^{\frac{s^2\mathbb{E}[G^2]}{2}}$. $\qquad\square$

REMARK 4.6 ρ-immune replication   From identity (4.74), we have

$$\mathbb{E}^{\mathbb{P}}[e^{p(S_T-S_0)}] = \mathbb{E}^{\mathbb{P}}[e^{p\rho\int_0^T a_t dB_t + \frac{q}{2}\langle S \rangle_T}]$$

$$= \mathbb{E}^{\mathbb{P}}[e^{\frac{q}{2}\langle S \rangle_T}\left(1 + p\rho\int_0^T a_t dB_t\right)] + O(\rho^2)$$

This implies that for all $\mathbb{P} \in \mathcal{M}^c(\rho, q)$,

$$\mathbb{E}^{\mathbb{P}}[e^{\frac{q}{2}\langle S \rangle_T}] = \sum_{i=\pm} \lambda_i \mathbb{E}^{\mu}[e^{p_i(S_T-S_0)}] + O(\rho^2) \equiv \mathrm{U}_{\mathrm{Carr-Lee}}(q) + O(\rho^2)$$

with ($p_\pm \equiv \pm\sqrt{q}$ here)

$$\sum_{i=\pm} \lambda_i = 1, \quad \sum_{i=\pm} \lambda_i p_i = 0$$

The resulting weights ($\lambda_- = \frac{1}{2}, \lambda_+ = \frac{1}{2}$) have been chosen such that the contribution of the term $\rho\mathbb{E}^{\mathbb{P}}[e^{\frac{q}{2}\langle S \rangle_T}\int_0^T a_t dB_t]$ cancels out. This was observed in [43]. At the first-order in ρ, this means that the price of $e^{\frac{q}{2}\langle S \rangle_T}$ is therefore model-independent within the class $\mathcal{M}^c(\rho, q)$. This gives a model-independent price for the payoff g_q (and therefore a general payoff g) at the first-order in ρ. In the present section, we extend this result and find robust model-independent bounds within the class $\mathcal{M}^c(\rho, q)$. As the case $\rho = 0$ is already covered in Section 3.4, we will assume that $\rho \neq 0$ in the rest of this section. $\qquad\square$

4.9.2 ρ-mixed SEP

For all $\mathbb{P} \in \mathcal{M}^c(\rho, \mu)$, we have from Lemma 4.1:

$$\mathbb{E}^{\mu}[\lambda(S_T)] = \mathbb{E}^{\mathbb{P}}[\lambda\left(\bar{S}_T + \gamma\langle\bar{S}\rangle_T Z\right)], \quad \gamma \equiv \frac{\sqrt{1-\rho^2}}{|\rho|}$$

where Z is a centered Gaussian variable $N(0,1)$. From DDS theorem, $\bar{S}_T = B_\tau$ where B is a Brownian motion and τ is a stopping time. This implies that

$$\mathbb{E}^{\mu}[\lambda(S_T)] = \mathbb{E}^{\mathbb{P}}[\lambda\left(B_\tau + \gamma\tau Z\right)]$$

Therefore finding an element $\mathbb{P} \in \mathcal{M}^c(\rho, \mu)$ maximizing $\mathbb{E}[g(\langle S \rangle_T)]$ is equivalent to finding a solution to a ρ-mixed SEP:

DEFINITION 4.5 ρ-Mixed SEP *Find a stopping time τ such that*

$$B_\tau + \gamma \tau Z \sim \mu, \quad \gamma \equiv \frac{\sqrt{1 - \rho^2}}{|\rho|}$$

Note that (± 1)-mixed SEP = SEP.

This defines a randomized SEP (see [98], Section 7.2.1 for some examples of explicit constructions - but not related to our present setting). We want to build an explicit example. By relying on our discussion on the Root solution, we are looking for a solution of the form

$$\tau_{\rho-\mathrm{R}} = \inf\{t > 0 \; : \; (t, B_t) \in R_\rho\}$$

with $R_\rho = \{(t, x) | t \geq r_\rho(x)\}$. As we impose that $\mathbb{B}_{\tau_{\rho-\mathrm{R}}} + \gamma \sqrt{\tau_{\rho-\mathrm{R}}} Z \sim \mu$:

$$\mu(x) = \mathbb{E}\left[\frac{e^{-\frac{(B_{\tau_{\rho-\mathrm{R}}} - x)^2}{2 r_\rho(B_{\tau_{\rho-\mathrm{R}}})\gamma^2}}}{\sqrt{2\pi r_\rho(B_{\tau_{\rho-\mathrm{R}}})\gamma^2}}\right]$$

Therefore, if we denote $\mu_\rho \equiv \mathrm{Law}(B_{\tau_{\rho-\mathrm{R}}})$, then

$$\mu(x) = \int \mu_\rho(dy) \frac{e^{-\frac{(x-y)^2}{2 r_\rho(y)\gamma^2}}}{\sqrt{2\pi r_\rho(y)\gamma^2}}, \quad \mu_\rho \equiv \mathrm{Law}(B_{\tau_{\rho-\mathrm{R}}}) \qquad (4.75)$$

As a conclusion, $\tau_{\rho-\mathrm{R}}$ is a solution to this randomized SEP if $r_\rho(\cdot)$ is a Root solution with μ replaced by μ_ρ. More precisely, from Proposition 4.9:

THEOREM 4.15

$\tau_{\rho-\mathrm{R}}$ *is a solution to the randomized SEP if $(r_\rho(\cdot), \mu_\rho(\cdot))$ solve the coupled equations*[1]

$$p(r_\rho(x)|x - S_0) = \int_\mathbb{R} 1_{r_\rho(x) \geq r_\rho(y)} p(r_\rho(x) - r_\rho(y)|x - y)\mu_\rho(dy)$$

$$\mu(x) = \int \mu_\rho(dy) \frac{e^{-\frac{(x-y)^2}{2 r_\rho(y)\gamma^2}}}{\sqrt{2\pi r_\rho(y)\gamma^2}}$$

---

[1]The author does not know how to prove that there is a solution for all μ. Of course, we know that there exists a solution for some μ: take a density μ_ρ, this will specify a unique r_ρ and therefore a density μ through equation (4.75).

4.9.3 Optimality

By restricting to the class $\mathcal{M}_q^c(\rho)$, the optimal bound can be computed analytically and compared to the Carr–Lee approach (see Remark 4.6). For use below, we define the 4 coefficients $(A_\pm, \lambda_\pm^*)$ which are uniquely fixed by the following 4 linear equations:

$$1 - \lambda_+^* e^{p_+ b_\pm} - \lambda_-^* e^{p_- b_\pm} =$$

$$A_+ \cos\left(\sqrt{\frac{q}{\rho^2}} b_\pm\right) + A_- \sin\left(\sqrt{\frac{q}{\rho^2}} b_\pm\right)$$

$$-\lambda_+^* p_+ e^{p_+ b_\pm} - \lambda_-^* p_- e^{p_- b_\pm} =$$

$$\sqrt{\frac{q}{\rho^2}} \left(-A_+ \sin\left(\sqrt{\frac{q}{\rho^2}} b_\pm\right) + A_- \cos\left(\sqrt{\frac{q}{\rho^2}} b_\pm\right)\right) \qquad (4.76)$$

where $b_\pm$ are the unique solutions of $(b_- < S_0 < b_+)$

$$\frac{e^{p_+ b_+}(S_0 - b_-) + e^{p_+ b_-}(b_+ - S_0)}{b_+ - b_-} = \mathbb{E}^\mu[e^{p_+ S_T}] \qquad (4.77)$$

$$\frac{e^{p_- b_+}(S_0 - b_-) + e^{p_- b_-}(b_+ - S_0)}{b_+ - b_-} = \mathbb{E}^\mu[e^{p_- S_T}] \qquad (4.78)$$

with $p_\pm \equiv \pm\sqrt{\frac{q}{1-\rho^2}}$. The 4 coefficients $(A_\pm, \lambda_\pm^*)$ can be found explicitly in terms of $b_\pm$. Their lengthy expressions are not reported.

In the particular case where μ is symmetric, i.e.,

$$\mathbb{E}^\mu[e^{p_+(S_T - S_0)}] = \mathbb{E}^\mu[e^{p_-(S_T - S_0)}]$$

we have

$$b_\pm = S_0 \pm \frac{1}{p_+} \cosh^{-1} \mathbb{E}^\mu[e^{p_+(S_T - S_0)}]$$

We set also

$$U(s) \equiv A_+ \cos\left(\sqrt{\frac{q}{\rho^2}} s\right) + A_- \sin\left(\sqrt{\frac{q}{\rho^2}} s\right)$$

THEOREM 4.16

(i) *Upper bound:*

$$\mathrm{MK}_q(\rho) = U(S_0) + \sum_{i=\pm} \lambda_i^* \mathbb{E}^\mu[e^{p_i S_T}]$$

(ii) *Optimality:* There exists $\mathbb{P}^* \in \mathcal{M}^c(\rho, q)$ such that $\mathrm{MK}_q(\rho) = \mathbb{E}^{\mathbb{P}^*}[e^{\frac{q}{2}\langle S\rangle_T}]$.

(ii)' *Comparison with Carr–Lee:*

$$\mathrm{MK}_q(\rho) = \mathrm{U}_{\mathrm{Carr-Lee}}(q) + O(\rho^2)$$

(iv) *Pathwise inequality:* $\forall\, \mathbb{P} \in \mathcal{M}^c(\rho)$,

$$e^{\frac{q}{2}\langle S\rangle_T} - \sum_{i=\pm} \lambda_i^* e^{p_i S_T} \leq U(S_0) + \int_0^T e^{\frac{q}{2}\langle S\rangle_t} \partial_s U(\bar{S}_t) d\bar{S}_t \tag{4.79}$$

$$+ \sum_{i=\pm} \lambda_i^* \int_0^T \left(e^{\frac{q}{2}\langle S\rangle_t - p_i(S_t - \bar{S}_t)} - 1 \right) d\mathbb{E}_t^{\mathbb{P}}[e^{p_i S_T}]$$

$$- \sum_{i=\pm} \lambda_i^* p_i \int_0^T e^{\frac{q}{2}\langle S\rangle_t - p_i(S_t - \bar{S}_t)} \mathbb{E}_t^{\mathbb{P}}[e^{p_i S_T}] d(S_t - \bar{S}_t)$$

(v) *Pathwise equality:*

$$e^{\frac{q}{2}\langle S\rangle_T} - \sum_{i=\pm} \lambda_i^* e^{p_i S_T} = U(S_0) + \sum_{i=\pm} \lambda_i^* \int_0^T \left(e^{\frac{q}{2}\langle S\rangle_t - p_i(S_t - \bar{S}_t)} - 1 \right) d\mathbb{E}_t^{\mathbb{P}^*}[e^{p_i S_T}]$$

$$- \sum_{i=\pm} \lambda_i^* p_i \int_0^T e^{\frac{q}{2}\langle S\rangle_t - p_i(S_t - \bar{S}_t)} \mathbb{E}_t^{\mathbb{P}^*}[e^{p_i S_T}] dS_t, \quad \mathbb{P}^* - \mathrm{a.s}$$

$$(4.80)$$

The pathwise inequality consists in a cash strategy $U(S_0) + \sum_{i=\pm} \lambda_i^* \mathbb{E}^\mu[e^{p_i S_T}]$, a delta hedging on S_t, a delta hedging on $e^{p_i S_T}$-payoffs, and a residual unhedged volatility risk proportional to $d(S_t - \bar{S}_t)$. This volatility risk vanishes under $\mathbb{P}^*$ and the pathwise inequality becomes an equality.

PROOF   **(i)** From the duality representation, we have

$$\mathrm{MK}_q(\rho) = \inf_{\lambda_\pm} \sup_{\mathbb{P} \in \mathcal{M}^c(\rho)} \mathbb{E}^{\mathbb{P}}[e^{\frac{q}{2}\langle S\rangle_T} - \sum_{i=\pm} \lambda_i e^{p_i S_T}] + \sum_{i=\pm} \lambda_i \mathbb{E}^\mu[e^{p_i S_T}]$$

$$\leq \sup_{\mathbb{P} \in \mathcal{M}^c(\rho)} \mathbb{E}^{\mathbb{P}}[e^{\frac{q}{2}\langle S\rangle_T} - \sum_{i=\pm} \lambda_i^* e^{p_i S_T}] + \sum_{i=\pm} \lambda_i^* \mathbb{E}^\mu[e^{p_i S_T}]$$

where $\lambda_\pm^*$ are defined above. By using Lemma 4.1, we obtain

$$\mathrm{MK}_q(\rho) \leq \sup_{\mathbb{P} \in \mathcal{M}^c} \mathbb{E}^{\mathbb{P}}[e^{\frac{q}{2\rho^2}\langle \bar{S}\rangle_T} \left(1 - \sum_{i=\pm} \lambda_i^* e^{p_i \bar{S}_T} \right)] + \sum_{i=\pm} \lambda_i^* \mathbb{E}^\mu[e^{p_i \bar{S}_T}]$$

where $\bar{S}$ is a martingale under $\mathbb{P} \in \mathcal{M}^c$ ($\bar{S}_0 = S_0$). By applying DDS theorem,

$$\mathrm{MK}_q(\rho) \leq u(S_0, 0) + \sum_{i=\pm} \lambda_i^* \mathbb{E}^\mu[e^{p_i S_T}]$$

where we have set

$$u(s,v) = \sup_\tau \mathbb{E}^{\mathbb{P}}[e^{\frac{q}{2\rho^2}\tau}\left(1 - \sum_{i=\pm}\lambda_i^* e^{p_i B_\tau}\right)|B_0 = s, \langle B\rangle_0 = v]$$

The function u is solution of the variational inequality:

$$\max\left(e^{\frac{q}{2\rho^2}v}\left(1 - \sum_{i=\pm}\lambda_i^* e^{p_i s}\right) - u(s,v), \partial_v u + \frac{1}{2}\partial_s^2 u\right) = 0$$

We take $u(s,v) = e^{\frac{q}{2\rho^2}v}U(s)$ and U satisfies

$$\max\left(1 - \sum_{i=\pm}\lambda_i^* e^{p_i s} - U(s), qU(s) + \rho^2\partial_s^2 U\right) = 0$$

A solution is $U(s) = A_+ \cos\left(\sqrt{\frac{q}{\rho^2}}s\right) + A_- \sin\left(\sqrt{\frac{q}{\rho^2}}s\right)$. From the expression of $(A_\pm, \lambda_\pm^*)$ listed above, we can check that U satisfies a C^1-smooth fit conditions at $s = b_\pm$, i.e.,

$$U(b_\pm) = 1 - \sum_{i=\pm}\lambda_i^* e^{p_i b_\pm}, \quad \partial_s U(b_\pm) = -\sum_{i=\pm}\lambda_i^* p_i e^{p_i b_\pm}$$

u defines then a supersolution of the above variational inequality. This implies

$$\mathrm{MK}_q(\rho) \le \overline{\mathrm{MK}}_q(\rho) \equiv U(S_0) + \sum_{i=\pm}\lambda_i^* \mathbb{E}^\mu[e^{p_i S_T}]$$

From Itô's lemma, we derive the pathwise inequality: For all $\mathbb{P} \in \mathcal{M}^c$,

$$U(S_0) + \int_0^T e^{\frac{q}{2\rho^2}\langle\bar S\rangle_t}\partial_s U(\bar S_t)d\bar S_t \ge e^{\frac{q}{2\rho^2}\langle\bar S\rangle_T}\left(1 - \sum_{i=\pm}\lambda_i^* e^{p_i \bar S_T}\right) \quad (4.81)$$

(ii) Set $\tau_{p\pm} \equiv \inf\{t > 0 : \mathbb{B}_t \notin [b_-, b_+]\}$ with $\mathbb{B}$ a Brownian motion, $\mathbb{B}_0 = S_0$, and $b_\pm$ specified by Equations (4.77,4.78). Then,

$$\mathbb{E}[e^{p_i \mathbb{B}_{\tau_{p\pm}} + \frac{q}{2\rho^2}\tau_{p\pm}}] = \mathbb{E}^\mu[e^{p_i S_T}], \quad i = \pm$$

We conclude the optimality of our bound by checking that

$$\mathbb{E}^{\mathbb{P}^*}[e^{\frac{q}{2}\langle S\rangle_t}] = \mathbb{E}[e^{\frac{q}{2\rho^2}\tau_{p\pm}}] = \overline{\mathrm{MK}}_q(\rho)$$

(ii)' See Remark 4.6.

(iii) We have the trivial identity:

$$e^{p\bar S_T + \frac{q}{2}\langle S\rangle_T} = \mathbb{E}[e^{p\bar S_T + \frac{q}{2}\langle S\rangle_T}] + \int_0^T d\mathbb{E}_t[e^{p\bar S_T + \frac{q}{2}\langle S\rangle_T}]$$

By using the mixing formula (4.73), we get for all $\mathbb{P} \in \mathcal{M}^c(\rho)$

$$e^{pS_T} = e^{p\bar{S}_T + \frac{q}{2}\langle S\rangle_T} - \int_0^T \left(e^{\frac{q}{2}\langle S\rangle_t - p(S_t - \bar{S}_t)} - 1\right) d\mathbb{E}_t[e^{pS_T}]$$

$$+ p \int_0^T e^{\frac{q}{2}\langle S\rangle_t - p(S_t - \bar{S}_t)} \mathbb{E}_t[e^{pS_T}] d(S_t - \bar{S}_t)$$

which implies that

$$e^{\frac{q}{2}\langle S\rangle_T} - \sum_{i=\pm} \lambda_i^* e^{p_i S_T} = e^{\frac{q}{2}\langle S\rangle_T} \left(1 - \sum_{i=\pm} \lambda_i^* e^{p_i \bar{S}_T}\right)$$

$$+ \sum_{i=\pm} \lambda_i^* \int_0^T \left(e^{\frac{q}{2}\langle S\rangle_t - p_i(S_t - \bar{S}_t)} - 1\right) d\mathbb{E}_t[e^{p_i S_T}]$$

$$- \sum_{i=\pm} \lambda_i^* p_i \int_0^T e^{\frac{q}{2}\langle S\rangle_t - p_i(S_t - \bar{S}_t)} \mathbb{E}_t[e^{p_i S_T}] d(S_t - \bar{S}_t)$$

By using inequality (4.81), we get our result (4.79).

(iv) The weak duality implies that the pathwise inequality becomes an equality for $\mathbb{P}^*$ defined above. This equality can be simplified. By using the mixing formula (4.74), we get

$$I \equiv \sum \lambda_i^* p_i e^{-p_i(S_t - \bar{S}_t)} \mathbb{E}_t^{\mathbb{P}^*}[e^{p_i \bar{S}_T}] = \sum_{i=\pm} \lambda_i^* p_i \mathbb{E}_t^{\mathbb{P}^*}[e^{p_i \bar{S}_t + \frac{q}{2}(\langle S\rangle_T - \langle S\rangle_t)}]$$

From DDS theorem, we have the identity

$$I \stackrel{\mathrm{DDS}}{=} \sum_{i=\pm} \lambda_i^* p_i e^{-\frac{q}{2}\langle S\rangle_t} \mathbb{E}_t^{\mathbb{P}^0}[e^{p_i \mathbb{B}_{\tau_{p_\pm}} + \frac{q}{2\rho^2}\tau_{p_\pm}}]$$

with $\mathbb{P}^0$ the Wiener measure. Then, we use that by construction (see Equation (4.76)) as $\mathbb{B}_{\tau_{p_\pm}} = b_\pm$ $\mathbb{P}^0$-a.s:

$$U'(\mathbb{B}_{\tau_{p_\pm}}) = -\sum_{i=\pm} \lambda_i^* p_i e^{p_i \mathbb{B}_{\tau_{p_\pm}}}, \quad \mathbb{P}^0 - \mathrm{a.s}$$

Therefore, we deduce that

$$I = -e^{-\frac{q}{2}\langle S\rangle_t} \mathbb{E}_t^{\mathbb{P}^0}[e^{\frac{q}{2\rho^2}\tau_{p_\pm}} U'(\mathbb{B}_{\tau_{p_\pm}})]$$

$$I \stackrel{\mathrm{DDS}}{=} -\mathbb{E}_t^{\mathbb{P}^*}[U'(\bar{S}_T) e^{\frac{q}{2}(\langle S\rangle_T - \langle S\rangle_t)}]$$

$$= -U'(\bar{S}_t)$$

where we have used in the last line that $U(\bar{S}_t)e^{\frac{q}{2}\langle S\rangle_t}$ is a martingale. This implies that the stochastic integral of the form $\int \ldots d\bar{S}_t$ in (4.79) disappears and hence our result (4.80). $\quad\square$

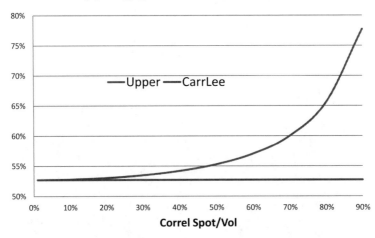

Figure 4.8: $U_q(\rho)$ versus $U_{Carr-Lee}$ as a function of $|\rho|$. μ is a Black–Scholes density with a volatility $\sigma_{BS} = 0.2$ and $T =$ one year.

Example 4.8 Volatility swap

In Figure 4.8, we have plotted $U_q(\rho)$ for $|\rho| \in [0,1]$. μ is here given by a Black–Scholes density with a volatility $\sigma_{BS} = 0.2$ and $T =$ one year. We have compared $U_q(\rho)$ with $U_{Carr-Lee}(q)$ which coincides for a Black–Scholes density with the Black–Scholes price $\exp(q\sigma_{BS}^2 T/2)$. We observe, as proved in Theorem 4.16 (ii)', that for small ρ

$$U_q(\rho) \approx U_{Carr-Lee}$$

In Figure 4.9, by using the following Laplace transform

$$\sqrt{V} = \frac{1}{2\sqrt{\pi}} \int_0^\infty \frac{1 - e^{-zV}}{z^{\frac{3}{2}}} dz$$

and the same data for μ, we have computed a lower bound for a volatility swap - See Equation (4.72). Note that in all numerical examples, $V_T = \langle \ln S \rangle_T$ (not $\langle S \rangle_T$) and Theorem 4.16 has been slightly modified to include this change. □

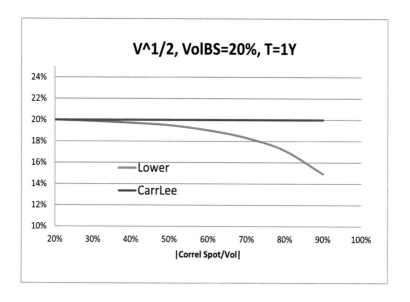

Figure 4.9: Lower bound for a volatility swap with payoff $\sqrt{\langle \ln S \rangle_T}$ versus Black–Scholes price as a function of $|\rho|$. μ is a Black–Scholes density with a volatility $\sigma_{\text{BS}} = 0.2$ and $T =$ one year.

References

[1] Acciaio, B., Beiglböck, M., Penkner, F., Schachermayer, W., Temme, J.: A trajectorial interpretation of Doob's martingale inequalities, *Ann. Appl. Probab.*, Volume 23, Number 4 (2013), pp. 1494–1505.

[2] Acciaio, B., Beiglböck, M., Penkner, F., Schachermayer, W.: A Model-free version of the fundamental theorem of asset pricing and the super-replication theorem, *Math. Finance*, 26(2):233–251, 2016.

[3] Albin, J. M. P.: A continuous non-Brownian motion martingale with Brownian motion martingale distributions, *Statistics and Probability Letters*, 78, 6, 682–686. 2008.

[4] Avellaneda, M., Paras, A.: Managing the volatility risk of portfolios of derivative securities: The Lagrangian uncertain volatility model, *Applied Mathematical Finance* Volume 3, 1996 - Issue 1.

[5] Avellaneda, M., Friedman, C., Holmes, R., Samperi, D.: Calibrating volatility surfaces via relative entropy minimisation, *Applied Mathematical Finance* Vol. 4, Issue 1,1997, Pages 37–64.

[6] Avellaneda, M.: Minimum-entropy calibration of asset-pricing models, *Int. J. Theor. App. Finance* 1(4) (1998) 447.

[7] Avellaneda, M., Buff, R., Friedman, C., Grandchamp, N., Kruk, L., Newman, J.: Weighted Monte Carlo: A new technique for calibrating asset-pricing models, *Int. J. Theor. App. Finance*, 2001, 4, pp. 1–29.

[8] Azéma, J., Yor, M.: Une solution simple au problème de Skorokhod, *Séminaire de Probabilités* XIII: 06, 90–115, LNM 721 (1979).

[9] Azéma, J.; Yor, M.: Le problème de Skorokhod: complément à l'exposé précédent, *Séminaire de Probabilités* XIII:06, 625–633, LNM 721 (1979).

[10] Bayraktar, E., Huang, Y.-J., Zhou, Z.: On hedging American options under model uncertainty, *SIAM J. Finan. Math.*, 6(1), 425–447.

[11] Bayraktar, E., Zhang, Y.: Fundamental theorem of asset pricing under transaction costs and model uncertainty, *Mathematics of Operations Research*, 41(3):1039–1054, 2016.

[12] Bayraktar, E., Zhou, Z.: On model-independent pricing/hedging using shortfall risk and quantiles, arXiv:1307.2493v1.

180   *References*

[13] Biagini, S., Bouchard, B., Kardaras, C., Nutz, M.: Robust fundamental theorem for continuous processes, *Math. Finance*, to appear, 2016.

[14] Beiglböck, M. , Cox, A., Huesmann, M., Perkowski, N., Prömel, D. J.: Pathwise super-replication via Vovk's outer measure, arXiv:1504.03644.

[15] Beiglböck, M., Cox, A., Huesmann, M.: Optimal transport and Skorokhod embedding, *Invent. Math.*, DOI 10.1007/s00222-016-0692-2 (2016).

[16] Beiglböck, M., Griessler, C.: An optimality principle with applications in optimal transport, arXiv:1404.7054v2.

[17] Beiglböck, M., Henry-Labordère, P., Penkner, F.: Model-independent bounds for option prices: A mass-transport approach, *Finance Stoch.*, July 2013, Volume 17, Issue 3, pp 477–501.

[18] Beiglböck, M., Juillet, N.: On a problem of optimal transport under marginal martingale constraints, *Ann. Probab.* Volume 44, Number 1 (2016), pp 42–106.

[19] Beiglböck, M., Nutz, M.: Martingale inequalities and deterministic counterparts, *Electron. J. Probab.* 19 (2014), no. 95, 1–15.

[20] Beiglböck, M., Nutz, M., Touzi, N.: Complete duality for martingale optimal transport on the line, to appear *Ann. Probab.*, arXiv:1507.00671.

[21] Beiglböck, M., Siorpaes, P.: Pathwise versions of the Burkholder–Davis–Gundy inequalities, *Bernoulli* Volume 21, Number 1 (2015), 360–373.

[22] Bentata, A., Cont, R.: Forward equations for option prices in semimartingale models, *Finance Stoch.* (2015) 19:617–651.

[23] Bergomi, L.: Smile dynamics II, *Risk Magazine*, (Oct. 2005).

[24] Bergomi, L.: *Stochastic Volatility Modeling*, Chapman & Hall/CRC Financial Mathematics Series (2016).

[25] Bouchard, B., Elie, R., Touzi, N.: Stochastic target problems with controlled loss, *SIAM Journal on Control and Optimization*, 48(5):3123–3150, 2009.

[26] Bouchard, B., Nutz, M.: Arbitrage and duality in nondominated discrete-time models, *The Annals of Applied Probability*, 25(2):823– 859, 2015.

[27] Bratelli, O., Robinson, D.W.: *Operator Algebras and Quantum Statistical Mechanics 1, Theoretical and Mathematical Physics*, Springer (1987).

[28] Breeden, D.T., Litzenberger, R.H.: Prices of state-contingent claims implicit in options prices, *J. Business*, 51, 621–651 (1978).

[29] Brenier, Y.: Décomposition polaire et réarrangement monotone des champs de vecteurs, *C. R. Acad. Sci. Paris* Série I Math., 305(19): 805–808, 1987.

[30] Brezis, H.: *Functional Analysis, Sobolev Spaces and Partial Differential Equations*, Universitext, Springer (2011).

[31] Brown, H., Hobson, D., Rogers, L.C.G.: Robust hedging of barrier options, *Math. Finance*, Vol. 11, No. 3, 285–314 (Jul. 2001).

[32] Brown, H., Hobson, D., Rogers, L.C.G.: The maximum maximum of a martingale constrained by an intermediate law, *Séminaire de Probabilités* XXXII, 250–263 (1998).

[33] Burkholder, D.: The best constant in the Davis inequality for the expectation of the martingale square function, *Trans. Amer. Soc.* 354 (2002).

[34] Campi, L., Laachir, I., Martini, C.: Change of numéraire in the two marginals martingale transport problem, *Finance Stoch.*, doi:10.1007/s00780-016-0322-2 (2016), arXiv:1406.6951.

[35] Campi, L., Martini, C.: On the support of extremal measures with given marginals: the countable case, arXiv:1607.07197.

[36] Carlier, G.: On a class of multidimensional optimal transport problems, *Journal of Convex Analysis*, Vol. 10 (2003), No. 2, 517-529.

[37] Carlier, G., Jimenez, C., Santambrogio, F.: Optimal transport with traffic congestion and Wardrop equilibria, *SIAM Journal on Control and Optimization*, 47(3):1330–1350, 2008.

[38] Carr, P., Madan, D.: Towards a theory of volatility trading, In R. Jarrow, editor, *Volatility*, pp 417–427. Risk Publications, 1998.

[39] Carr, P.: Local variance Gamma option pricing model, presentation, *IBCI Conference*, Paris (April 2009).

[40] Carr, P.: Options on maxima, drawdown, trading gains, and local time, preprint.
http://www.math.csi.cuny.edu/probability/Notebook/skorohod3.pdf.

[41] Carr, P., Chou, A.: Breaking barriers: Static hedging of barrier securities, *Risk Magazine* 10 (1997) 139–145.

[42] Carr, P. , Geman, H., Madan, D.B., Yor, M.: From local volatility to local Lévy models, *Quantitative Finance*, Volume 4, Issue 5, 2004.

[43] Carr, P., Lee, R.: Realized volatility and variance: Options via swaps, *Risk Magazine*, May 2007.

[44] Carr, P., Lee, R.: Robust replication of volatility derivatives, preprint (2009). http://www.math.uchicago.edu/ rl/rrvd.pdf.

[45] Carr, P., Lee, R.: Hedging variance options on continuous semimartingales, *Finance Stoch.*, vol 14 issue 2 (2010), 179–207.

[46] Carr, C., Lee, R.: Put-call symmetry: Extensions and applications, *Math. Finance*, Volume 19, Issue 4, October 2009, 523–560.

[47] Carraro, P., El Karoui, N., Oblój, J.: On Azéma–Yor processes, their optimal properties and the Bachelier drawdown equation, *Annals of Probability* 40(1): 372–400 (2012).

[48] Claisse, J., Guo, G., Henry-Labordère, P.: Robust hedging of options on local time, arXiv:1511.07230.

[49] Cont, R., Deguest, R.: Equity correlations implied by index options: estimation and model uncertainty analysis, *Math. Finance*, Vol. 23, No. 3 (July 2013), 496–530.

[50] Cox, A., Hobson, D., Oblój, J.: Pathwise inequalities for local time: Application to Skorokhod embeddings and optimal stopping, *Ann. Appl. Probab.* 18(5), 1870–1896 (2008).

[51] Cox, A., Oblój, J.: Robust pricing and hedging of double no-touch options, *Finance Stoch.*, 15(3):573–605, 2011.

[52] Cox, A., Oblój, J., Touzi, N.: The root solution to the multi-marginal embedding problem: An optimal stopping and time-reversal approach, ArXiv:1505.03169.

[53] Cox, A., Wang, J.: Root's barrier: Construction, optimality and applications to variance options, *Ann. Appl. Probab.*, Volume 23, Number 3 (2013), 859–894.

[54] Cox, A., Wang, J.: Optimal robust bounds for variance options, arXiv:1308.4363.

[55] Davis, B.: On the L^p norms of stochastic integrals and other martingales, *Duke Math. J.* Volume 43, Number 4 (1976), 697-704.

[56] Davis, M.H.A., Hobson, D.: The range of traded option prices, *Math. Finance*, 17(1):1–14, 2007.

[57] De March, H., Touzi, N.: Irreducible convex paving for decomposition of multi-dimensional martingale transport plans, arXiv:1702.08298.

[58] De Marco, S., Henry-Labordère, P.: Linking Vanillas and VIX options: A constrained martingale optimal transport, *SIAM J. Finan. Math.*, 6(1), 1171–1194.

[59] Delbaen, F., Schachermayer, W.: *The Mathematics of Arbitrage*, Springer Finance (2006).

[60] Delbaen, F., Grandits, P., Rheinlander, T., Samperi, D., Schweizer, M., Stricker, C.: Exponential hedging and entropic penalties, *Math. Finance*, Volume 12, Issue 2, pp 99-123, April 2002.

[61] Deng, S., Tan, X.: Duality in nondominated discrete-time models for American options, arXiv:1604.05517.

[62] Denis, L. and Martini, C.: A Theoretical framework for the pricing of contingent claims in the presence of model uncertainty, *Annals of Applied Probability* 16 (2), 827-852 (2006).

[63] Dolinsky, Y., Soner, H.M.: Robust hedging and martingale optimal transport in continuous time, *Probability Theory and Related Fields* October 2014, Volume 160, Issue 1, pp 391–427.

[64] Dolinsky, Y., Soner, H.M.: Robust hedging under proportional transaction costs, *Finance Stoch.* April 2014, Volume 18, Issue 2, pp 327–347.

[65] Dolinsky, Y., Soner, H.M.: Martingale optimal transport in the Skorokhod space, *Stochastic Processes and Their Applications*, Volume 126, Issue 1, January 2016, pp 312–313.

[66] Dubins, L.E., Schwarz, G.: On extremal martingale distributions, *Proc. Fifth Berkeley Symp. on Math. Statist. and Prob.*, Vol. 2, Pt. 1 (Univ. of Calif. Press, 1967), 295–299.

[67] Dupire, B.: *Arbitrage Pricing with Stochastic Volatility, Derivatives Pricing*, The Classic Collection (Editor Peter Carr), Risk Books, 2004.

[68] Dupire, B.: Pricing with a smile, *Risk Magazine* 7, 18–20 (1994).

[69] Dupire, B.: Applications of the Root Solution of the Skorokhod embedding problem in Finance, slides, Bloomberg (2007).

[70] El Karoui, N., Rouge, R.: Pricing via utility maximization and entropy, *Math. Finance*, Volume 10, Issue 2, pp 259–276, April 2000.

[71] Fahim, A., Huang, Y.-J.: Model-independent superhedging under portfolio constraints, *Finance Stoch.*, January 2016, Volume 20, Issue 1, pp 51–81.

[72] Fan, J.Y., Hamza, K., Klebaner, F.C.: Mimicking self-similar processes, *Bernoulli* Volume 21, Number 3 (2015), 1341–1360.

[73] Föllmer, H.: Calcul d'Itô sans probabilités, *Séminaire de probabilités*, XV: 09, 143–150, Lecture Notes in Mathematics 850 (1981).

[74] Föllmer, H., Schied, A.: *Stochastic Finance: An Introduction in Discrete Time, de Gruyter Studies in Mathematics* (Book 27), Walter de Gruyter edition (November 24, 2004).

[75] Föllmer, H., Leukert, P.: Quantile Hedging, *Finance Stoch.*, Vol. 3, Issue 3, May 1999.

[76] Fukasawa, M.: Volatility derivatives and model-free implied leverage, *IJTAF* 17 (2014), no.1, 1450002.

[77] Galichon, A., Henry-Labordère, P., Touzi, N.: A stochastic control approach to no-arbitrage bounds given marginals, with an application to Lookback options, *Ann. Appl. Probab.* Volume 24, Number 1 (2014), 312–336.

[78] Gangbo, W., Święch, A.: Optimal maps for the multidimensional Monge–Kantorovich prooblem, *Comm. Pure Appl. Math.*, 51 (1) : 23–45, 1998.

[79] Gangbo, W., McCann, R.: The geometry of optimal transport, *Acta. Math.* 177(2):113–161, 2006.

[80] Gassiat, P., Mijatovi, A., Oberhauser, H.: An integral equation for Root's barrier and the generation of Brownian increments, *Ann. Appl. Probab.*, Vol. 25, No 4 (2015), 2039–2065.

[81] Gassiat, P., Oberhauser, H., dos Reis, G.: Root's barrier, viscosity solutions of obstacle problems and reflected FBSDEs, *Stochastic Processes and Their Applications*, 2015, Vol. 125, No. 12, pp 4601–4631.

[82] Ghoussoub, N., Kim, Y-H., Lim, T.: Structure of optimal martingale transport in general dimensions, arXiv:1508.01806.

[83] Ghoussoub, N., Kim, Y-H., Lim, T.: Optimal Skorokhod embedding for radially symmetric marginals in general dimension, preprint, 2016.

[84] Glasserman, P., Yang, L.: Bounding wrong-way risk in CVA calculation, preprint (2015), http://ssrn.com/abstract=2648495.

[85] Gangbo, W., McCann, R.: The geometry of optimal transport, *Acta Math.*, 177(2):113–161, 1996.

[86] Guo, G., Tan, X., Touzi, N.: Optimal Skorokhod embedding under finitely-many marginal constraints, *SIAM J. Control Optim.*, 54(4), 2174–2201.

[87] Guyon, J., Henry-Labordère, P.: *Nonlinear Option Pricing*, Financial Mathematics Series CRC, Chapman & Hall (440 p.), chief-editor: M.A.H. Dempster, D. Madan, R. Cont (2013).

[88] Guyon, J., Menegaux, R., Nutz, M.: Bounds for VIX futures given S&P 500 smiles, ssrn.com/abstract=2840890.

[89] Hambly, B., Mariapragassam, M., Reisinger, C.: A forward equation for barrier options under the Brunick & Shreve Markovian projection, *Quantitative Finance*, Volume 16, 2016 - Issue 6.

[90] Hamza, K., Klebaner, F.C.: A family of non-Gaussian martingales with Gaussian marginals, *J. Appl. Math. and Stochastic Analysis*, Article Id 92723, 2007.

[91] Henry-Labordère, P.: Calibration of local stochastic volatility models to market smiles, *Risk Magazine* (Sep. 2009).

[92] Henry-Labordère, P.: Optimal transport, geometry and numerical algorithms for nonlinear PDEs , https://tel.archives-ouvertes.fr/tel-01088419.

[93] Henry-Labordère, P. : Automated option pricing: Numerical method, *IJTAF*, Volume 16, Issue 08, Dec. 2013.

[94] Henry-Labordère, P., Oblój, J., Spoida, P., Touzi, N.: The maximum maximum of a martingale with given marginals, *Ann. Appl. Probab.* Volume 26, Number 1 (2016), 1–44.

[95] Henry-Labordère, P., Touzi, N.: An explicit martingale version of Brenier's theorem, *Finance Stoch.*, July 2016, Volume 20, Issue 3, pp 635–668.

[96] Henry-Labordère, P., Touzi, N., Tan, X.: An Explicit martingale version of Brenier's theorem with infinitely many marginals, *Stochastic Processes and Their Applications* 126:9, 2800–2834.

[97] Hirsch, F., Profeta, C., Roynette, B., Yor, M.: Constructing self-similar martingales via two Skorokhod embeddings, *Séminaire de Probabilités* XLIII, Lecture Notes in Mathematics 2011, pp 451–503.

[98] Hirsch, F., Profeta, C., Roynette, B., Yor, M.: *Peacocks and Associated Martingales, with Explicit Constructions*, Springer (2011).

[99] Hirsch, F., Roynette, B.: A new proof of Kellerer's theorem, *ESAIM: Probability and Statistics*, Vol.16, Sept. 2012, pp 48–60.

[100] Hobson, D.: Robust hedging of the lookback option, *Finance Stoch.*, 2, pp 329–347, 1998

[101] Hobson, D.: *The Skorokhod Embedding Problem and Model-Independent Bounds for Option Prices*, In Paris-Princeton Lectures on Mathematical Finance 2010, volume 2003 of Lecture Notes in Math., pp 267–318. Springer, Berlin, 2011.

[102] Hobson, D.: Fake exponential Brownian motion, *Statistics and Probability Letters*, Volume 83, Issue 10, October 2013, pp 2386–2390.

[103] Hobson, D.: Model free hedging, Presentation, Bachelier World Congress Brussels, June 2014.

[104] Hobson, D., Klimmek, M. : Robust price bounds for the forward starting straddle, *Finance Stoch.* January 2015, Volume 19, Issue 1, pp 189–214.

[105] Hobson, D., Neuberger, A.: Robust bounds for forward start options, *Math. Finance*, Vol 22, Issue 1, pp 31–56 (Jan. 2012).

[106] Hobson, D., Neuberger, A.: On the value of being American, arXiv:1604.02269.

[107] Hodges, S., Neuberger, A.: Optimal replication of contingent claims under transactions costs, *The Review of Futures Markets*, Vol 8, No 2, 1989.

[108] Hou, Z., Oblój, J.: On robust pricing-hedging duality in continuous time, arXiv:1503.02822, preprint (2015).

[109] Jacod, J, Yor, M.: Etude des solutions extrémales et réprésentation intégrale des solutions pour certains probèmes de martingales, *Z. Wahr. Verw. Gebiete* 38 (1977), No 2, p. 83–152.

[110] Kallblad, S., Tan, X., Touzi, N.: Optimal Skorokhod embedding given full marginals and Azéma–Yor peacocks, to appear *Ann. Appl. Probab.*, arXiv:1503.00500.

[111] Kellerer, H.G.: Markov-Komposition und eine anwendung auf martingale, *Math. Ann.* 198:99–122, 1972.

[112] Korman, J. , McCann, R.: Optimal transport with capacity constraints, *Trans. Amer. Math. Soc.* 367 (2015), 1501–1521.

[113] Madan, D.B., Yor, M.: Making Markov martingales meet marginals: with explicit constructions, *Bernoulli*, 8(4):509–536, 2002.

[114] McCann, R.J.: A convexity principle for interacting gases, *Advances in Mathematics* (128):153–179, 1997.

[115] Méléard, S.: *Asymptotic Behaviour of Some Interacting Particle Systems; McKean–Vlasov and Boltzmann Models, in Probabilistic Models for Non-Linear Partial Differential Equations*, Lecture Notes in Mathematics 1627, Springer-Verlag (1996).

[116] Nutz, M.: Pathwise construction of stochastic integrals, *Electronic Communications in Probability*, Vol. 17, No. 24, pp. 1–7, 2012.

[117] Neuberger, A.: Volatility trading, London Business School working paper, 1990.

[118] Oblój, J.: The Skorokhod embedding problem and its offspring, *Probability Surveys*, 1: 321–392 (2004).

[119] Oblój, J.: A complete characterization of local martingales which are functions of Brownian motion and its maximum, *Bernoulli*, 12: 955–969, 2006.

[120] Oblój, J., Siorpaes, P.: Structure of martingale transports in finite dimensions, arXiv:1702.08433.

[121] Oblój, J., Spoida, P.: An Iterated Azéma-Yor type embedding for finitely many marginals, *Annals of Probability*, to appear, available at arXiv:1304.0368.

[122] Oleszkiewicz, K.: On fake Brownian motions, *Statistics and Probability Letters*, 78, 1251–1254, 2008.

[123] Osekowski, A.: Sharp maximal inequalities for the martingale square bracket, *Stochastics: International Journal of Probability and Stochastic Processes*, Volume 82, Issue 6, 2010.

[124] Pass, B.: Uniqueness and Monge solutions in the multimarginal optimal transport problem, *SIAM J. Math. Anal.*, 43(6):2758–2775, 2011.

[125] Pass, B.: On a class of optimal transport problems with infinitely many marginals, *SIAM J. Math. Anal.*, 45(4), 2557–2575.

[126] Peccati, G., Yor, M.: Burkholder's submartingales from a stochastic calculus perspective, *Illinois J. Math.* Volume 52, Number 3 (2008), 815–824.

[127] Perkins, E.: The Cereteli-Davis solution to the H1-embedding problem and an optimal embedding in Brownian motion, in Seminar on Stochastic Processes, 1985 (Gainesville, Fla., 1985). *Progr. Probab. Statist.* 12 172–223. Birkhäuser, Boston.

[128] Peskir, G., Shiryaev, A. N.: Optimal Stopping and Free-Boundary Problems, Lectures in Mathematics, Birkhäuser Basel (2006).

[129] Pham, H.: On some recent aspects of stochastic control and their applications, *Probab. Surveys* Volume 2 (2005), 506–549.

[130] Revuz, D., Yor, M.: *Continuous Martingales and Brownian Motion, Grundlehren der Mathematischen Wissenschaften*, Springer (1999).

[131] Root, D.H.: The existence of certain stopping times on Brownian motion, *Ann. Math. Statist.* Volume 40, Number 2 (1969), 715–718.

[132] Rost, H.: The stopping distributions of a Markov process, *Inventiones Mathematicae* 14 (1971), pp. 1–16.

[133] Sawyer, N.: SGCIB launches timer options, *Risk Magazine*, July 2007.

[134] Stebegg, F.: Model-independent pricing of Asian options via optimal martingale transport, arXiv:1412.1429, Dec. 2014.

[135] Strassen, V.: The existence of probability measures with given marginals, *Ann. Math. Statist.*, 36:423–439 (1965).

[136] Soner, H.M., Touzi, N., Zhang, J.: Quasi-sure analysis through agregation, *Electron. J. Probab.* Volume 16 (2011), paper no. 67, 1844–1879.

[137] Vallois, P.: Le problème de Skorokhod sur $\mathbb{R}$: une approche avec le temps local, Séminaire de probabilités de Strasbourg, 17 (1983), p. 227–239.

[138] Vallois, P.: Quelques inégalités avec le temps local en zéro du mouvement Brownien, *Stochastic Processes and Their Applications* 41 (1992) 117-155.

[139] Villani, C.: Topics in optimal transport, *Graduate Studies in Mathematics AMS*, Vol 58.

[140] Vovk, V.: Rough paths in idealized financial markets, *Lith. Math. J.*, 51(2):274–285, 2011.

[141] Vovk, V.: Continuous-time trading and the emergence of probability, *Finance Stoch.* 16(4):561–609, 2012.

[142] Vovk, V.: Itô calculus without probability in idealized financial markets, *Lith. Math. J.*, 55(2):270–290, 2015.

[143] Yor, M.: Le mouvement Brownien: une martingale exceptionnelle et néanmoins générique, Lecons de mathématiques d'aujourd'hui, Vol. 3, p 103–138, Edition Cassini (2007).

[144] Zaev, D.: On the Monge–Kantorovich problem with additional linear constraints, *Mathematical Notes*, 2015, 98:5, 725741, arXiv:1404.4962, 2014.

Index

Admissible portfolio, 113
Arbitrage opportunity, 9
Arbitrage-free price, 9
Attainable payoff, 12
Azéma–Yor martingale, 94
Azéma–Yor solution, 134

Bachelier implied volatility, 106
Bachelier model, 107
Bass's construction, 117
Bauer maximum principle, 14
Bid-ask price, 70
Binomial model, 11
Black–Scholes formula, 22
Brenier's solution, 34
Burkholder–Davis–Gundy inequality, 165
Buyer's super-replication price, 9

c-concave transform, 29
c-cyclical monotonicity, 61
Call options, 1
Canonical space, 89
Complete model, 13
Concave envelope, 52
Conditionally symmetric martingale, 106
Convex displacement, 62
Convex order, 48
Copula, 31
Corridor variance swap, 92
Covariance options, 92
CVA, 35

Dambis–Dubins–Schwarz theorem, 100
Doob's inequality, 163
Drawdown constraint, 95

Equivalent martingale measure, 21
Equivalent probabilities, 6
Exponential utility, 18
Extremal point, 13

Fake Brownian motion, 127
Fat tails, 3
Feasible convex set, 10
Forward-start Azéma–Yor solution, 140
Forward-start option, 45
Fréchet–Hoeffding solution, 31

Gibbs density, 85

Hamilton–Jacobi equation, 43
Hamilton–Jacobi–Bellman equation, 51
Hopf–Lax's formula, 44

Implied volatility, 26

Kellerer's theorem, 48
Kullback–Leibler distance, 84

Left-monotone martingale, 54
Leveraged Exchange Trading Fund (LETF), 96
Lévy's identity, 110
Local stochastic volatility model, 121
Local Bergomi model, 47
Local Lévy's model, 122
Local time, 98
Local variance Gamma model, 118
Local volatility model, 120
Log-normal density, 2
Log-swap, 39
Lookback options, 93

Markov MOT, 65
Martingale Brenier's solution, 58
Martingale convex displacement, 64
Martingale convex interpolation, 63
Martingale Fréchet–Hoeffding solution, 54, 123
Martingale Monge–Kantorovich duality, 47
Martingale Spence–Mirrlees condition, 60
McKean–Vlasov SDE, 121
Mean-variance hedging, 14
Minimal martingale entropy probability, 18
Mixed SEP, 171
Model-independent arbitrage-free price, 70
Model-independent arbitrage, 69
Monge's cost, 51
Monge's solution, 34
Monge–Ampère equation, 36
Monge–Kantorovich duality, 28
MOT formulation with two periods, 44
MOT in continuous-time, 115
Multi-marginal MOT, 65
Multi-marginals Azéma–Yor solution, 143

Ocone's martingale, 105
Options on local time, 97
Options on variance, 96
OT formulation with 2 assets, 27
OT formulation with multi assets, 42

Pathwise integration, 113
Perkins solution, 152
Prokhorov's theorem, 11
Pure martingale, 106

Quantile hedging, 16

Range timer options, 101
Relative entropy, 18

Relatively compact, 11
Replication, 12
Risk-neutral probability, 7
Robust quantile hedging, 41, 67
Root's solution, 146
Running maximum, 93

Self-financing portfolio, 20
Seller's super-replication price, 6
Short/long position, 5
Singular probabilities, 114
Skorokhod embedding problem (SEP), 130
Spence–Mirrlees condition, 31, 39
Strictly monotone function of order 2, 42

Tanaka's formula, 98
Tight family, 11
Timer option, 98
Transaction cost, 70
Trinomial model, 18
Two-factor Bergomi model, 47

Utility indifference buyer's price, 17

Vallois solution, 156
Vanilla options, 25
Variance swap, 66
VIX future, 79
VIX options, 79
Volatility swap, 166

Wasserstein distance, 121
Weak/Strong duality, 8